AF763606

Agriculture in the People's Republic of China

Leslie T. C. Kuo

The Praeger Special Studies program—utilizing the most modern and efficient book production techniques and a selective worldwide distribution network—makes available to the academic, government, and business communities significant, timely research in U.S. and international economic, social, and political development.

Agriculture in the People's Republic of China

Structural Changes and Technical Transformation

PRAEGER SPECIAL STUDIES IN INTERNATIONAL ECONOMICS AND DEVELOPMENT

Praeger Publishers New York Washington London

Library of Congress Cataloging in Publication Data

Kuo, Leslie Tse-chiu, 1914-
Agriculture in the People's Republic of China.

(Praeger special studies in international economics and development)
Includes bibliographical references and index.
1. Agriculture—China. 2. Agriculture and state—China.
I. Title.
S471.C6K85 338.1'0951 75-36412
ISBN 0-275-05870-0

PRAEGER PUBLISHERS
111 Fourth Avenue, New York, N.Y. 10003, U.S.A.

Published in the United States of America in 1976
by Praeger Publishers, Inc.

WITHDRAWN
University of Bristol

UNIVERSITY
OF BRISTOL
LIBRARY

All rights reserved

© 1976 by Praeger Publishers, Inc.

Printed in the United States of America

PREFACE

A country's agricultural productivity depends on a number of interrelated factors. Because of the diversity of political, economic, social, physical, and other conditions, the factors involved vary greatly from country to country. In the People's Republic of China (PRC), a socialist system of ownership and management and a progressive program for technological improvement are considered as two essentials to increase agricultural productivity. These are generally referred to as the "socialist transformation" and the "technical transformation" of agriculture. Mao Tse-tung stated in 1955: "We are now carrying out a revolution not only in the social system, the change from private to public ownership, but also in technology, the change from handicraft to large-scale modern machine production, and the two revolutions are interconnected."

Although this principle is generally agreed upon among the Communist Chinese leaders, the programming of the two types of transformation has been a matter of dispute between Mao Tse-tung and Liu Shao-ch'i. Mao was of the opinion that early cooperativization would provide a more rational and efficient organization of labor and use of land, would make it easier for the state to collect agricultural products, and would speed up capital accumulation and industrialization—that only the large-scale farming achieved through cooperativization could create suitable conditions for massive application of modern technology. For this reason, he argued, the socialist transformation of agriculture should play a more important role than the technical transformation during the First Five-Year Plan (1953-57) and the Second Five-Year Plan (1958-62), and the two should proceed at the same pace during the Third Five-Year Plan. Mao estimated that although the socialist transformation would reach its final stage at the end of the Third Five-Year Plan, it would take four or five five-year plans, or from twenty to twenty-five years, to complete the technical transformation.[1]

Liu Shao-ch'i, on the other hand, advocated the postponement of cooperativization until industry was ready to carry out capital construction and to supply chemical fertilizers, agricultural machinery, and other necessities to the countryside. In his opinion it was especially necessary to bring about mechanization before carrying out cooperativization, because collective farms cannot be consolidated without mechanization, which depends on industrialization.[2]

The socialist transformation has been carried out in several stages. At the outset land ownership was redistributed according to a "land reform" program. Under a law promulgated in June 1950, some 700 million mou of land and large numbers of draft animals, farm tools, and other property were redistributed to farmers.[3] Later, farms were reorganized with an increasing degree of cooperativization through the establishment of the mutual-aid teams, next the primary agricultural producers' cooperatives (APCs), and then the advanced APCs. Finally, when the Leap Forward campaign was launched in 1958, farms in the PRC were placed under the management of the communes.

Meanwhile, numerous measures were taken to innovate farming techniques as a means of improving agricultural production. Increasing attention was given to scientific research, training of technical personnel, and popularization of improved farming methods. Mao Tse-tung repeatedly urged the Chinese Communist Party (CCP) to strive for the technical transformation of agriculture, although his primary priority was cooperativization. He frequently stated, "The fundamental way out for agriculture lies in mechanization."

The present volume covers both forms of transformation that the Chinese authorities consider essential to the agricultural development in China. Part I is a general discussion of the socialist transformation of agriculture, reviewing and interpreting such topics as the organization and management of farms, ownership of the means of production, and distribution of income and farm produce at various stages of the institutional transitions (land reform, cooperativization, and communalization). Part II is an updated and expanded version of The Technical Transformation of Agriculture in Communist China, published in 1971. In this part the overall policies, guiding principles, programs, and methods for the technological innovation of agriculture adopted during the first quarter-century of the Communist regime will be examined. This examination will be followed by a discussion of the major technical measures for improving agricultural production. In discussing each technical measure, primary attention will be given to the planning and execution of the programs rather than to the details of agricultural technology. In each case the situation before the Communist take-over, the highlights of the programs carried out by the new regime, and the major accomplishments and problems will be observed. On the basis of such a perspective, the progress of the technical transformation of agriculture will be evaluated and an attempt will be made to determine, inter alia, the extent to which the technical innovation has helped agricultural productivity in the PRC and the prospects for the next few decades.

The evaluation of Chinese agriculture is a complex undertaking. First, China is a large and varied country, extending across some 35 degrees of latitude and with extreme variations in elevation, precipitation, and soil. There are also great differences in economic and social conditions among parts of the country. Consequently many types of agriculture are practiced. Second, agricultural technology is itself not an entity but a complex of fields. Even the simplest approach involves an examination of a number of major topics. The present study is only a first step toward bringing into focus a general picture of institutional changes and technical innovations and their effects on agricultural production for the first twenty-five years of the Chinese Communist regime. Additional studies of a specialized regional or local nature will be needed for understanding the problems more thoroughly.

A further difficulty in making the present study is the lack of precise, dependable information. Most of the materials were compiled from Communist sources. In addition, the author has canvassed materials published in the United States, Taiwan, Hong Kong, and Japan for the information not found in the Chinese Communist sources. Since publishing is a state monopoly under the Communist regime, a considerable portion of the information supplied by the authorities is inevitably for propaganda purposes. Although every caution has been taken in selecting and checking the materials, some facts and figures quoted in this study should be accepted with reservation.

During the Cultural Revolution (1966-69) and for several years afterward, publishing in the PRC was virtually suspended and materials for research were practically nonexistent. It was not until early 1972 that new publications, including some on agriculture, began to appear, mainly because of the gradual improvement of the general political situation. The resumption of publishing, together with the normalization of relations with the United States and other countries, has motivated many libraries to expand their Chinese collections. The inflow of Chinese publications has been facilitated by the visit to the United States in October 1973 of a delegation of librarians from China. During this visit the matter of promoting exchange of publications between American and Chinese libraries was informally discussed. Since then new books and periodicals published in the PRC have slowly but increasingly arrived in the United States.

In preparing this enlarged volume the latest materials available have been incorporated into the original compilation, and some of the previous interpretations and analyses have been changed accordingly. In addition to the data for the post-Cultural Revolution period, the reader will find some new facts and figures for the earlier years, which had not been made public during the preparation of the first edition but have become available since then. A trip to the PRC in the summer of 1975 made it possible for the author to make on-the-spot observations and to acquire additional materials.

Despite the increasing availability of research materials, there remain a number of gaps to be filled, and discussions on some topics in this study must be brief because of the lack of data. It should also be emphasized that this volume is designed to be primarily a reference book on Chinese agriculture with a broad coverage. As such, it contains only a limited amount of analytical assessment. The author hopes that from the discussions it is at least possible to obtain a fairly clear idea of what has actually taken place in Chinese agriculture in the twenty-five years since the inauguration of the PRC in 1949.

NOTES

1. Mao Tse-tung, "On the Question of Agricultural Cooperativization," report delivered at a meeting of the secretaries of the provincial, municipal, and district committees of the Chinese Communist Party on July 31, 1955, in Jen-min Shou-ts'e [People's Handbook] (Peking, 1956), pp. 80-86.

2. "Hold High the Great Red Banner of Mao Tse-tung's Thought and Completely Knock Down China's Krushchev," Nung-ts'un Ch'ing-nien [Rural Youth], Shanghai, Nos. 17-18, September 10, 1967, and September 25, 1967; translated in Current Background, Hong Kong, No. 847, February 16, 1968, pp. 1-29.

3. Lias Lu-yen, "A great Victory in the Land Reform Movement During the Past Three Years," in Committee for the Promotion of China's Trade, ed., San-nien-lai Hsin-chung-kuo Ching-chi ti Ch'eng-chiu [New China's Economic Accomplishments in the Past Three Years] (Peking, 1952), pp. 111-18.

ACKNOWLEDGMENTS

The compilation of the materials for the present study began in the early 1950s. A grant voted by the Joint Committee on Contemporary China of the Social Science Research Council and the American Council of Learned Societies enabled the author to take a leave of absence in 1963 and 1964 from his regular duties at the U.S. Department of Agriculture and to devote his full time to the completion of the first draft of the manuscript. Since then the draft has been drastically revised and updated according to the latest data available to cover the first quarter-century of the PRC.

A major portion of the materials used in this study come from the National Agricultural Library of the Department of Agriculture, which maintains one of the largest collections of publications on Chinese agriculture. Of special importance to this study are some 1,000 monographs, 150 periodicals, and several leading Communist Chinese newspapers in the collection. Also frequently used are Chinese materials from Taiwain and English materials on the PRC, including Current Scene, China News Analysis, Far Eastern Economic Review, and Union Research Service; materials compiled by the U.S. consulate general in Hong Kong (Survey of People's Republic of China Press, Current Background, and Selections from People's Republic of China Magazines); Agricultural information on Mainland China, compiled by the American agricultural attache in Hong Kong; translations by the Joint Publications Research Service in Washington, D.C.; and releases of the New China News Agency and the China News Service in Peking. The author is most grateful to Dr. Foster E. Mohrhardt, former director of the National Agricultural Library, and Dr. Richard A. Farley, the incumbent, for the facilities the library provided, which expedited this study's progress enormously.

In addition to drawing upon many sources of materials, the author has consulted a number of specialists, including several in the U.S. Department of Agriculture. Special mention should be made of the advice of the late Dr. John Lossing Buck, for many years head of the department of agricultural economics at the University of Nanking, whose monumental Land Utilization in China and many other writings on Chinese agriculture are among the major classics in this field. Dr. Buck painstakingly read the manuscript of Part II (technical transformation) and made invaluable suggestions. The author also wishes to thank Dr. Ardron B. Lewis, formerly of the Agricultural Development Council, New York, for his thorough review of the manuscript of Part I (socialist transformation); and

Professor Alexander Eckstein of the University of Michigan, who kindly reviewed the initial draft of the study. He deeply appreciates the keen interest and encouragement of Dr. Don Paarlberg, director of agricultural economics, U. S. Department of Agriculture. For the technical terms used in this study, the Agricultural Research Service of the U. S. Department of Agriculture has frequently been consulted. Dr. John L. Creech, director of the National Arboretum, Washington, D. C., has given helpful advice, particularly for the chapter on plant breeding.

The completion of the present study would not have been possible without the unceasing inspiration and encouragement of the author's wife, Hsu-hua, who has participated in practically all stages of the present project. A chemist, she has been particularly helpful in the preparation of the chapters on fertilization and plant protection.

Needless to say, the author is solely responsible for the views and interpretations presented and for the errors that inevitably will be found in this volume.

CONTENTS

Chapter | Page

LIST OF TABLES

LIST OF MAPS

CHINESE
WEIGHTS AND MEASURES

Weight

1 tan (picul) = 0. 055 short ton = 0. 050 metric ton
1 chin (catty) = 1. 102 pounds = 0. 500 kilogram
1 liang (tael) = 1. 102 ounces = 0. 031 kilogram

Length

1 li = 0. 311 mile = 0. 500 kilometer
1 chang = 3. 645 yards = 3. 333 meters
1 ch'ih = 1. 094 feet = 0. 333 meter
1 ts'un = 1. 312 inches = 0. 333 decimeter

Area

1 ch'ing = 16. 474 acres = 6. 667 hectares
1 mou = 0. 164 acre = 0. 067 hectare

LIST OF ABBREVIATIONS

APC	Agricultural Producers' Cooperative
ARL	Agrarian Reform Law
CCP	Chinese Communist Party
FBIS	Foreign Broadcast Information Service
JMJP	*Jen-min Jih-pao* [People's Daily], Peking
JPRS	Joint Publications Research Service, Washington, D. C.
NCNA	New China News Agency, Peking
NPAD	National Program for Agricultural Development
PCC	Production-Construction Corps
PLA	People's Liberation Army
PRC	People's Republic of China

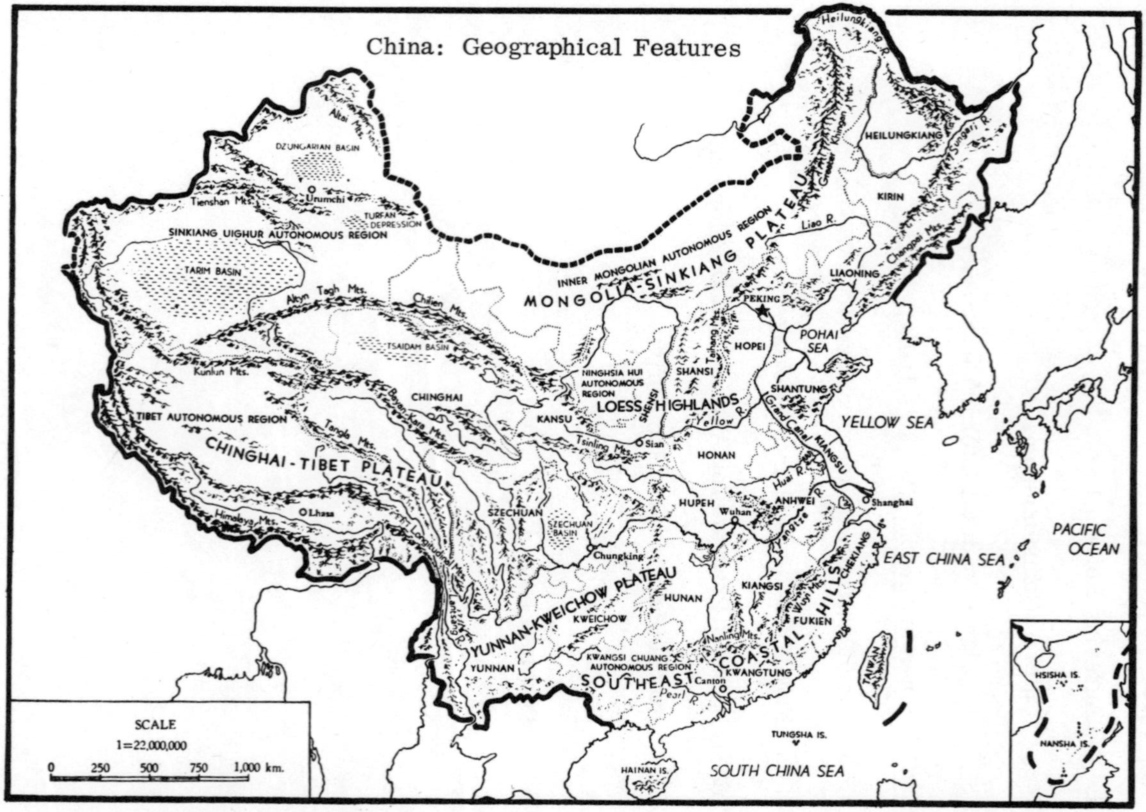

Source: China Reconstructs

PART I

SOCIALIST TRANSFORMATION OF AGRICULTURE

INTRODUCTION

The socialist transformation of agriculture concerns the institutional changes of the farming system, involving such problems as the organization and management of farms, ownership of the means of production, distribution of income and farm produce, power to make plans for production, and development of the productive forces. At the outset of the Communist Chinese regime, land ownership was redistributed according to the Agrarian Reform Law (ARL) of 1950. The completion of land distribution in 1952 was followed by the organization of the masses of peasants toward gradual collectivization, each step marking a greater departure from traditional land tenure and management. First, farms throughout the country were organized into mutual-aid teams, which contained the rudiments of socialism and were later reorganized into the semi-socialist primary APCs. By the end of 1956, more than 100 million farm households had been further reorganized into 750,000 advanced APCs, which were fully socialist. Finally, "in the upsurge of agricultural production and socialist construction" in the second half of 1958, the advanced cooperatives were merged into more than 70,000 people's communes, which were a modified form of collective. *

* Sue Hsing, (Wo-kuo Nung-yeh Ti She-hui Chu-i Tao-lu [The Socialist Road of Our Country's Agriculture] (Peking: People's Press, 1976).

CHAPTER

1

LAND REFORM

Over the years the problem of farm tenancy in China has been debated on political and economic grounds. Most of the earlier studies were based on a vague egalitarian notion rather than on a specific justification. Studies of the problem in the 1930s produced more dependable information. Some of these works indicated that the extent of farm tenancy in China had often been exaggerated and that the degenerating economic condition in China's countryside should be attributed to overpopulation, uneconomical fragmentation of land, and backward farming methods rather than to "the excessive concentration of land ownership." The studies maintained that there were no possible direct economic effects of land redistribution in terms of increased agricultural production.

John Lossing Buck's study, Land Utilization in China, conducted in 1929-33, indicated that on 16,876 farms in 22 provinces, 54 percent of farmers were owners; 29 percent, part owners; and 17 percent, tenants.[1] It also showed that 28.7 percent of the farm area was rented. Estimates made by the National Agricultural Research Bureau of the Nationalist government, based on yearly reports of several thousand voluntary crop reporters in 22 provinces, indicated that in 1931-37, 46 percent of the farmers were owners; 25 percent, part owners; and 29 percent, tenants.

According to a report of the China-United States Agricultural Mission, published in 1947, about 30 percent of the farmland in China was rented. The mission also estimated that about 50 percent of the Chinese farmers owned the land they operated; more than one-fourth owned some of the land they operated and rented additional land; and nearly one-fourth rented all the land they operated.[2] Compared with other countries, the prevalence of tenancy in China in the 1940s was no greater. For instance, England had about 75 percent of its farmers as tenants; the United States, 42 percent; Germany, 25 percent; and Japan, 27 percent.[3]

Several surveys conducted in the 1930s and the 1940s indicated that many tenants were better farm business operators than owners, and that many farmers preferred to be tenants. Some of the surveys showed that farm tenancy could be developed as a mutually beneficial partnership between the landlord and the tenant. There were, of course, certain evils in farm tenancy, such as excessive rent charged by the landlords and absentee landlordism, which was prevalent in certain localities, especially the Lower Yangtze and the Canton delta areas. The principal problem, according to the surveys, was preventing evils from developing within the system. For solving this problem the China-United States Agricultural Mission recommended setting rents fair to both the landlords and the tenants; providing funds and facilities for the purchase of land by individual tenants through an improved farm credit system; and conducting studies to determine tenancy abuses and land ownership conditions and farm leasing practices that needed correction.[4] The Nationalist government made sporadic attempts at rent reduction and the demonstration of "tenant land purchase" projects in a few isolated regions, but these had just been started when the Communists took over.[5]

The Chinese Communists, on the other hand, regarded China's existing land ownership system in the 1920s as one of feudalism and exploitation. In his "Report on the Question of Agrarian Reform" to the Second Session of the National Committee of the Chinese People's Political Consultative Conference on June 14, 1950, Liu Shao-ch'i explained:

> Landlords and rich peasants, who constitute less than 10 percent of the rural population, possess approximately from 70 to 80 percent of the land and brutally exploit the peasants by means of their land. Poor peasants, farm laborers, middle peasants, and others, however, who make up 90 percent of the rural population, possess in all only 20 to 30 percent of the land. They toil all the year round but can hardly have a full belly and warm back. . . . Herein lies the basic reason why our nation has become theobject of aggression and oppression and has become impoverished and backward. This also constitutes the principal obstacle to our nation's democratization, industrialization, independence, unification, and prosperity. Unless we change this situation, the victory of the Chinese people's revolution cannot be consolidated, the productive forces in the rural areas cannot be set free, the industrialization of New China cannot be realized, and the people cannot enjoy the fundamental gains of the victory of the revolution.[6]

The purpose of land reform in the PRC was, therefore, "to set free the rural productive forces, develop agricultural production and thus pave the way for New China's industrialization."[7] A major feature of the program was the confiscation of the land owned by the landlords for distribution to the landless or land-poor peasants. The Communists believed that the transfer of land ownership to the poor peasants and farm laborers would not only elevate their economic status but also would influence them to replace the old system of social values with an entirely new one. They also believed that land reform would release a part of the productive forces that was not fully and efficiently used under the old system, and that increased productivity in rural areas would provide a basis for improving the living standard of the peasants.

The policy development stage of the program actually went back to the prewar Kiangsi-Soviet phase of the early 1930s. Two important documents, How to Analyze Class Status in the Countryside and Decisions Concerning Some Problems Arising from Agrarian Reform, were promulgated in 1933 by the Democratic Central Government in Juichin, Kiangsi Province, "to ensure a correct solution of agrarian problems."[8] But until the latter part of 1947, only piecemeal policy statements and party directives related to land reform had been issued. It was not until October 10 of that year that the CCP promulgated the "Summary Land Law," which contained rules and regulations in a more compact and legalistic form. The 16 articles of the law defined the expropriations (without compensation) of the lands of the large landowners and the confiscation of the lands of temples, churches, and schools; the division of the land among the people; the abolition of all previous debts and contracts; and the right and freedom of the new owners to buy and sell. It stated the main object of the expropriation to be the reduction in the power of the enemies of the regime, the large landowners, but not that of the "middle landowners."

The Agrarian Reform Law (ARL) promulgated on June 30, 1950, significantly modified the "Summary Land Law" of October 10, 1947. The old law called for the requisition of the surplus rural properties of rich peasants. The new law stipulated that land owned by rich peasants and worked by them or their hired laborers was not to be touched. Rich peasants also could retain small portions of land that they rented to others. In addition, people engaged in nonagrarian occupations, such as workers or professional people, who owned and rented out small parcels of land were not to be classified as landlords and were allowed to keep and to rent out such parcels. Thus the new changes involved principally the redistribution of land from landlords, and to a lesser extent from rich peasants, to poor and landless peasants. The dual purpose of these changes was to encourage the early restoration of stability and agricultural production and to isolate the landlords.

The operational organs that made up the distributory machinery for land reform included the land reform committees, the people's tribunals, and the peasants' associations. The land reform committees were set up at various levels from the great administrative areas down to the hsien, and were composed of persons elected by people's representative conferences or appointed by the people's governments of a higher level. These committees were responsible for directing and handling all matters concerning land reform. Those in each greater administrative area and in each province were primarily policy-making and decision-making authorities, while those at the hsien level were concerned purely with implementation.[9]

Article 32 of the ARL stipulated that people's tribunals be formed at the hsien level, with the principal responsibility of maintaining "revolutionary order" in the countryside under land reform. Their constitution was provided in the "General Regulations Governing the Organization of the People's Tribunals," promulgated by the Government Administrative Council on July 20, 1950. The presiding judges, deputy judges, and half the jurors were to be members of the hsien governments, and half the jurors were to be elected by the conferences of people's representatives or by mass organizations.[10] The tribunals were to travel to different places to try and punish, according to law, "the hated despotic elements who have committed heinous crimes, whom the masses of the people demand to be brought to justice, and all such persons who resist or violate the provisions of the Agrarian Reform Law and decrees."[11]

Shortly after the promulgation of the ARL "General Regulations Governing the Organization of Peasants' Associations" were adopted by the Government Administrative Council on July 14, 1950. These peasants' associations were mass organizations formed on a voluntary basis and designated as the legal executive organizations for reforming the agrarian system. Their functions were (1) to unite farm laborers, poor peasants, middle peasants, and all anti-feudal elements in the rural areas, so as to carry out social reforms step by step, and to protect the interests of the peasants in accordance with the policy, laws, and decrees of the people's government; (2) to organize the peasants for production, to set up rural cooperatives, to develop agriculture and subsidiary occupations, and to improve the living conditions of the peasants; and (3) to safeguard the political rights of the peasants, to raise their political and cultural level, and to take part in the construction of a state of people's democracy.[12]

The peasants' associations were based on the principle of "democratic centralism" (subordination of individual members to the organization, subordination of minorities to the majority, and the subordination of lower organizations to higher organizations). All measures of land reform were executed through the peasants' associations, which provided a nucleus of peasantry who had to be educated to meet the responsibilities of executing the agrarian reform laws. They formed a natural core of future leadership in the villages. In the process of implementing land reform laws, they came together for common action that must have formed a good ground for the development of the mutual-aid teams and the cooperatives.[13]

The peasants' associations were formed in each village, hsiang, hsien, and province. The village associations determined the class status of each person (the basis of the Chinese land reform), the land to be acquired, and the manner of distribution. Their decisions were subject to the approval of the higher bodies and subsequent ratification by government. There was also a right of appeal for each person affected by the land reform. In practice, however, the decisions of the village associations were generally upheld. Even when a village association had erred, it was persuaded to modify its decision rather than having the decision overruled. The peasant masses were thus organized and established. They were made to feel that they were part of government and shared in its authority. In the process the government and the Communist party, which were already close to the people, came still nearer to the peasants and thus were able to bring about a transformation of the Chinese agrarian economy.[14]

The first Chinese Communist peasants' association was organized in the early 1920s in Haifeng, Kwangtung Province. Later peasants' associations (in some cases called "peasants' unions" or "poor peasants' corps") were set up by various Chinese Communist leaders, including Mao Tse-tung, to pave the way for armed struggles of the Chinese Communists in the rural areas. By 1927 some 13 million peasants had been absorbed into the peasants' associations, which were the most popular form or organization adopted by the Chinese Communists during the Land Revolution in 1927-37, the Sino-Japanese War in 1937-45, the insurrection period in 1946-49, and the land reform period in 1949-52. In certain areas during certain periods, the associations served as basic executive or administrative offices for the Chinese Communist movement.

Following the attainment of the basic goals of the land reform in 1949-52, peasants' associations on all levels that had played an important role in the large-scale land reform began to lapse into oblivion and become disorganized. In 1953-63, the Chinese Communist party carried out its rural programs through its core organizations and exercised direct control over the peasants. Helping the party workers were "activists," who worked for no definite organization. There were no peasants' organizations in any form until the establishment of the poor and lower-middle peasants' associations during the Rural Socialist Education Movement (1963-65).[15]

In general the land reform was carried out in several stages: agitation and propaganda, division of classes, confiscation of land and other property for redistribution, and the issuance of title deeds.

A work team was first dispatched to initiate a propaganda and indoctrination campaign to arouse the "class consciousness" of the peasants. The activists were urged to form peasants' associations. The second step was to organize mass meetings to accuse and punish the evil landlords and despots, and to define the class status of the villagers. The next stage was the confiscation of the surplus property and the surplus land of landlords and certain rich peasants, and the distribution of these properties among the poor peasants and farm laborers. A victory celebration concluded the process with more propaganda and the issue of new title deeds.

"Decisions Concerning the Differentiation of Class Status in the Countryside," adopted by the Government Administrative Council on August 4, 1950, classified China's productive forces in the rural areas into five groups: landlords, rich peasants, middle peasants, poor peasants, and workers (including farm laborers). The criteria were as follows: A landlord lived on rent and other nonlabor income. A rich peasant worked on the farm but also hired labor. A middle peasant worked on his own land with no hired labor or rented land. Poor peasants and farm laborers had no land or a small amount of land and farm implements, and worked as tenants or hired hands.

In distributing land the general principle followed was that every family in the village received land. The land that the tiller acquired through land distribution, plus his own existing holdings, if any, was slightly and suitably more than his own holding before land distribution. Landlords were given a share equal to what the tiller acquired, so that they could make their living by their own labor. Land owned by rich and middle peasants and cultivated by themselves or by hired labor and their other properties were, in most cases, left untouched. Room was left for local discretion concerning land that rich and middle peasants leased to others. After redistribution all the families in the village were thus owners of land and generally no family held more than double the average holding. Thus the land distribution considerably narrowed the differences between rich, middle, and poor peasants.[16]

During the 1920s and the early 1930s the Chinese Communists had experimented with land reform. In some of the Communist-occupied areas, the Red Army confiscated land and redistributed it to the landless. After the outbreak of the Sino-Japanese War in 1937, the united front was formed with the Nationalist government, and the Communists adopted a moderate rent-reduction program. During the early stage of the Civil War of 1945-49, more progressive land reform programs were carried out in northern and northeastern China. In the winter of 1949-50, the people's government extended its land reform activities to the suburbs of the cities and a number of other areas in northern China. By the end of May 1950, agrarian reform in China had been completed, or mostly completed, in an area with a rural population of about 145 million. The total population of the area was about 160 million. There was still a section with a rural population of about 264 million where agrarian reform had not been carried out; its total population was about 310 million.[17]

When the ARL was promulgated on June 30, 1950, most of the rural population north of the Yangtze River had already completed land reform. On the third anniversary of the PRC (October 1, 1952), slightly over two years since the promulgation of the ARL, more than 90 percent of the rural population in China was reported to have accomplished land reform; and only 30 million were yet to do so in either the winter of 1952 or the spring of 1953.[18]

By the end of 1952, some 700 million mou (120 million acres), or 44 percent of China's total cultivated area of about 1,650 million mou (274 million acres), 30 million draft animals, 39 million agricultural implements, 5 million tons of foodstuffs, surplus houses containing about 38 million rooms, and other properties owned by landlords had been confiscated and redistributed to some 300 million poverty-stricken peasants who had little or no land and other means of production. In addition, the peasants no longer had to pay rent to the landlords, which formerly amounted to as many as 70,000 million chin (35 million metric tons) of grain each year.[19]

From the political viewpoint the Chinese Communist land reform was undoubtedly a success. The immediate political objective—gaining the support of poor peasants—was achieved. The landlords as a class were virtually eliminated, and the traditional land tenure system was completely eliminated. As a whole the program was implemented smoothly and there were only a few cases of organized mass opposition, mostly in the peripheral provinces. The economic impact of the program on agricultural production, however, was difficult to gauge because there were other factors involved, including the restoration of law and order, increase in farmland capital construction (especially irrigation and flood control), and economic rehabilitation in general. Lack of reliable data on these factors makes it impossible to determine the relationship between the postreform structure of ownership and productivity changes.

In any case, China's land reform did serve as a first step in abolishing "feudal production relations," which were regarded as impeding the development of productive forces, and as an integral part of the socialist transformation of agriculture. To that end, the fundamental objective of the land reform movement was attained.

NOTES

1. John Lossing Buck, Land Utilization in China, University of Nanking, 1937, pp. 192-99.

2. Report of the China-United States Agricultural Mission (Washington, D. C. : Office of Foreign Agricultural Relations, U. S. Department of Agriculture, May 1947), p. 17.

3. John Lossing Buck, "Farm Tenancy in China," Economic Facts, Department of Agricultural Economics, University of Nanking, Chengtu, No. 33, June 1944, and No. 34, July 1945, pp. 469-97.

4. Report of the China-United States Agricultural Mission, pp. 59-60.

5. Comments by Franklin L. Ho on John Lossing Buck, "The Agricultural Economy of China," in Paul K. T. Sih, ed., The Strenuous Decade: China's Nation-building Efforts, 1927-1937 (New York: St. John's University Press, 1970), pp. 194-203.

6. Hsin-hua Yueh-pao (New China Monthly) Peking, Vol. II, No. 3, July 1950, pp. 492-96.

7. The Agrarian Reform Law of the People's Republic of China, Article 1.

8. The Agrarian Reform Law of the People's Republic of China, and Other Relevant Documents (Peking: Foreign Languages Press, 1959), pp. 17-50.

9. Agrarian Reform Law . . . , Article 28, Section 5.

10. Jen-min Shou-ts'e [People's Handbook] (Shanghai: Ta Kung Pao Press, 1951), pp. 47-48, Part "cheng."

11. Agrarian Reform Law, Article 32, Section 5.

12. The Agrarian Reform Law of the People's Republic of China and Other Relevant Documents, pp. 51-57.

13. Planning Commission, Report of the Indian Delegation to China on Agrarian Co-operatives (New Delhi, 1957), pp. 32-33.

14. Ibid., p. 87.

15. Li Ming-hua, "Peasants' Associations and Class Struggle in Rural Areas in China Mainland," Issues and Studies, Taipei, Vol. No. 7, April 1965, pp. 46-52.

16. The Agrarian Reform Law of the People's Republic of China, Article 31; Planning Commission, op. cit., pp. 31-32; and State Statistical Bureau, Ten Great Years (Peking, 1960), p. 34.

17. Liu Shao-ch'i "Report on the Question of Agrarian Reform," presented to the Second Session of the National Committee of the Chinese People's Political Consultative Conference on June 14, 1950, Hsin-hua Yueh-pao, Peking, Vol. II, No. 3, July 1950, pp. 492-96.

18. Liao Lu-yen, "The Great Victory of the Land Reform Movement in the Past Three Years," Hsin-hua Yueh-pao, No. 36, October 1952, pp. 13-15; John Wong, Land Reform in the People's Republic of China (New York: Praeger, 1973), p. 128.

19. Planning Commission, loc. cit.; State Statistical Bureau, loc. cit.

CHAPTER

2

COOPERATIVIZATION

Land redistribution led some to believe that the new leaders in China were not Communists but "land reformers." Actually, however, as early as November 29, 1943, Mao Tse-tung, in speaking to a group of model laborers in the Communist-occupied "Shensi-Kansu-Ninghsia border areas," made it clear that the objective of land reform was the attainment of a socialist society, and urged them to organize:

> For thousands of years the peasant masses have maintained individual holdings wherein each family, each household, represented a separate economic unit. Such dispersed individual production is the economic foundation of a fuedal regime, and it dooms peasants to perpetual poverty. The only way to remedy this situation is to introduce gradual collectivization, and the only method of collectivization, as Lenin taught us, is through cooperation.[1]

Mao Tse-tung's view was reaffirmed in The Common Program, adopted on September 29, 1949, at the People's Political Consultative Conference, which stipulated:

> In all areas where agrarian reform has been thoroughly carried out, . . . it shall be the central task of the People's Government to organize the peasants and all manpower available for agriculture, for the development of agricultural production and side occupations. The People's Government shall also guide the peasants step by step in the organization of various forms of mutual aid in labor and cooperation in production, according to the voluntary principle and the principle of mutual benefit.[2]

The situation in the rural areas after land reform was such that the feudal land ownership by the landlord class had been abolished, the principle of "land to the tiller" had been put into practice, and productive forces had been released. Agricultural production was, however, still carried out in a scattered and individual way, with the single household the productive unit. Such a small-scale individual economy was capable of only very limited growth. It could not meet the ever-rising requirements of the nation's socialist industrialization for marketable grains and industrial raw materials any more than it could meet the demands of the expansion of the national economy or fulfill the need of the masses of the peasants to raise their living standards.

At the same time the Communist authorities were concerned that this kind of economy was perpetuating capitalist elements. After the agrarian reform most of the poor peasants and farm laborers had risen to the level of middle peasants, who therefore constituted the most important stratum of the rural population. In some areas where land reform had been completed comparatively early, the spontaneous forces of capitalism had already started to grow in the countryside: the sale of land, exploitation through the hiring of labor, commercial speculation, and usury had appeared. It was feared that if such a rural economy were allowed to develop, a small number of peasants would become rich peasants, the majority would go bankrupt, and the masses of peasants would continue to live a poor and hard life. All this convincingly demonstrated that should capitalism be allowed to spread unchecked, the peasants would be deprived of all their gains from the land reform.

It also was maintained that China's peasants had, under the leadership of the Chinese Communist party, become an awakened and organized group. The party, therefore, had a strong organizational base in the countryside, and the peasants were ready to take the socialist road. Immediately after the completion of land reform, the Chinese Communist party put forward the task of agricultural cooperativization by organizing the following groups: (1) seasonal mutual-aid teams, the simplest form of collective labor; (2) year-round mutual-aid teams, which had a division of labor and assignment of definite jobs on the basis of collective labor, and a certain amount of common Property; (3) primary (semisocialist) agricultural, in which the members pooled their land as shares and there was a single management and a greater amount of common property; and (4) advanced APCs (collective farms), which were completely socialist and were based on the peasants' collective ownership of land and other means of production.

MUTUAL-AID TEAMS

Following the completion of the land reform program in 1952, agricultural mutual-aid teams were organized in 1953 as a simple form of collective labor. Peasants joining the mutual-aid teams worked together but retained ownership of their land, draft animals, farm tools, and other means of agricultural production. The farm produce of each piece of land went to the family that owned it. Thus the peasants aided each other by exchanging labor days as they worked together. The mutual-aid team was only the "rudiment of socialism," because the means of production were still privately owned and land management was not yet centralized. The peasants banded together only for collective labor.

At the beginning the organization of mutual-aid teams was temporary and seasonal. They were later developed into year-round groups in which there was, on the basis of working together, a certain degree of division of labor and a small amount of collectively owned property. Temporary, seasonal mutual-aid teams usually consisted of three to five or, in rare cases, as many as seven peasant households helping one another with manpower and animal power. When it was time for large-scale plowing, harvesting, and transplanting of seedlings, they worked together; during the slack season they worked individually. There were no hard-and-fast rules regarding the record and norm of work one person did for another. There was no accurate bookkeeping, and only a rough estimate was made of the labor and services contributed by each member. This type of mutual-aid team was already well-known to Chinese farmers and was based on the peasants' traditional practices. It played a positive role in the countryside after the agrarian reform in restoring production and in overcoming the peasants' shortage of means of production. It also helped to inculcate ideas about collectivism among the peasants.

Larger in scale, the regular, year-round mutual-aid team generally comprised 10 to 20, or even a whole village of some 30, peasant households. All the work was done by the peasants together, but it was recorded and a norm fixed for each household. The year-round mutual-aid team usually had a leader and an accountant (if one could be found). Its members retained ownership of the means of production, but some decisions began to be made on a collective basis and some teams accumulated capital equipment. Such teams usually did not last long, because many of them were almost eligible to become primary APCs. Some of the year-round mutual-aid teams provided the opportunity for technical improvements in farming and for a rational division of labor. The year-round mutual-aid teams used a more accurate method for computing the labor days each had put in, and mapped out their rough or detailed production plans.[3]

Although operated under the system of private ownership of the means of production, the mutual-aid teams as a whole were reported to be effective in organizing the peasants into collective labor. They led to more production because of better farming obtained through teamwork, and ensured better tillage, rational use of draft animals, and the introduction of better agricultural implements that individual farmers were unable to secure. Specialization of labor and the capacity to meet urgent demands of work helped to increase production by enabling agricultural operations to be done intensively and at the proper time. Mutual-aid teams also helped to develop subsidiary organizations, such as transport work and labor on public works projects.[4]

The "Decisions on Agricultural Production for 1952," issued by the Administrative Council of the central people's government, provided that mutual-aid and cooperative organizations of various types should be developed throughout the country in accordance with local conditions. The country was urged to have 80 or 90 percent of its productive forces organized within two or three years.[5] By the end of August 1952, more than 40 percent of the peasants throughout the country had participated in the agricultural mutual-aid or cooperative organizations. In the areas that the Communists took over earlier, 70 to 80 percent of the peasants had taken part in such organizations.[6]

In reviewing the history of the agricultural cooperative movement in China, Mao Tse-tung stated in a report delivered at a conference of secretaries of provincial, municipal, and autonomous region committees of the CCP on July 31, 1955:

> In the twenty-two years of revolutionary wars preceding the founding of the People's Republic of China, our party had already acquired experience in guiding the peasants, after land reform, to form agricultural producers' mutual-aid organizations containing some rudiments of socialism. During that period, there had been mutual-aid working groups and plowing teams in Kiangsi Province, work-exchange teams in northern Shensi and mutual-aid teams in various places in northern, eastern and northeastern China. In isolated cases, agricultural producers' cooperatives of a semi-socialist or socialist character also came into being. . . . But such cooperatives were not being widely promoted then.[7]

Mutual-aid working groups and plowing teams, based on individual farming, were formed during the Civil War by peasants in the Communist-occupied areas to facilitate production through a better organization of labor power. On the principle of voluntary participation and mutual benefit, the members did an equal amount of work for each other, or if one could not give another as much help as he received, he made up the difference in cash. Besides helping each other the members of the groups gave preferential treatment to the families of Red Army soldiers and worked for bereaved old people without any pay except for meals. These measures of mutual aid reportedly were of great help to agricultural production; and since they were carried out on a reasonable basis, they won the warm support of the peasants.[8]

PRIMARY APCS

The primary APCs were the principal form of agricultural organization put into practice from 1953 to 1955. Under their management a small part of the income was distributed as "land payment" in proportion to the amount of land pooled by the members, while the greater part of the income was distributed as "work payment" to the members doing collective labor according to the quantity and quality of work done. Thus the system was an expression of the continued private ownership of land and, at the same time, embodied partial implementation of the socialist principle of "to each according to his work." Households in a primary APC still held their land privately but it was farmed in common, in accordance with the plans made by a central management. It was maintained that since the primary cooperatives carried out unified management of land, they could overcome the contradiction between collective labor and individual management, which had existed in mutual-aid teams. As a result, it was believed, they could make rational use of land and could more efficiently deploy the manpower and plan the farm work.

In joining the primary APCs the peasants pooled their land as shares and were paid according to the amount of work contributed and the amount of land contributed. A primary APC usually consisted of 10 to 20 households. Some comprised 40 to 50 households and others 80 to 100 households. Such cooperatives tilled an average of 30 hectares, and some cultivated as many as 200 to 250 hectares.

Each primary cooperative was formed on a permanent basis and had a chairman, committees, and accountants. In theory, members were supposed to have some influence in making major farm decisions because they constituted parts of the "congress of members," which approved important policy issues. In practice, however, the chairman exercised great influence and made day-to-day decisions.

The primary APCs were established in 1950 on an experimental basis. At the end of that year, there were only 18 such cooperatives throughout the country. After December 1951, when the CCP Central Committee promulgated a draft "Resolution of Mutual Aid and

Cooperation in Agricultural Production" for trial enforcement by the party committees at all levels, the number of primary APCs rose to 3,634 by the end of 1952. In February 1953 the CCP Central Committee formally promulgated the resolution after making certain revisions; and in keeping with the "general line of the Party during the transition period" set forth by Mao Tse-tung in October of the same year, the number of primary cooperatives increased to 15,053.

Following the promulgation by the CCP Central Committee of the "Decision on the Development of Agricultural Producers' Cooperatives" in mid-January 1954, the number of primary APCs rose rapidly. At the end of 1954 it was 114,165. However, Mao Tse-tung was not satisfied. In his report "On the Question of Agricultural Cooperation," delivered on July 31, 1955, at a conference of secretaries of provincial, municipal, and autonomous region committees of the CCP, he called hesitant cadres to task and branded them "women with bound feet." After this a great number of cooperatives were established with much fanfare. In November 1955 the Standing Committee of the National People's Congress adopted the "Draft of Model Regulations for Agricultural Producers' Cooperatives." By the end of 1955, 633,742 APCs had been set up and 529 had already been made advanced APCs.[9]

The primary APC was semisocialist because, on the one hand, it made unified use of land and rational use of farm tools, carried out collective labor, practiced distribution if income and farm produce according to work, and had a fair amount of common property; on the other hand, the members retained private ownership of land and other means of production, and received dividends on their land shares and payments for the pooling of their farm tools and draft animals. This kind of APC was a transitional form on the road to the fully socialist cooperative with collective ownership.[10]

ADVANCED APCS

In 1955 and 1956, many primary APCs were reorganized to become advanced APCs, which consisted of 100 or more households. A chairman and several committees were elected from the "congress of members" of the cooperative. To promote production and administrative efficiency, the labor force was divided into production brigades, which were made up of production teams. Farm management and other important decisions were made by the chairman of the cooperative, who was heavily influenced by the party.

In the advanced APCs practically all land was turned over to common ownership, but each member was permitted to use a small piece of "self-retained land" (private plots), which amounted to no more than 5 percent of the average acreage tilled by a member of the cooperative, for growing vegetables, fruits, and other horticultural crops. Land payment was abolished. The privately owned draft animals, farm tools, and other means of agricultural production were pooled at their money value and became the collective property of the cooperative. After setting a part of the value of these means of production against the cost of membership shares, the rest of their value was repaid in installments.

Income of the advanced cooperative was distributed among the members under the "labor day" work-payment system, on the basis of the socialist principle of "to each according to his work and more income for those who work more." A labor day represented the time and the amount of labor spent completing the norm set for each kind of work and for attaining certain standards required for each kind of work, including the technical standard, the arduousness of the labor process, and the importance of the work to the entire production process. At the end of the agricultural year, the gross income of the cooperative was totaled; deductions were made for production and management costs, agricultural taxes, reserve funds, public welfare funds, and other expenditures; and the resulting net income was divided by the total number of labor days credited to all the farmers and staff of the cooperative in order to determine the monetary value of a single labor day. The net income was then distributed to individual farm families according to the number of labor days they had earned.

In addition to distributing income, the advanced cooperatives implemented an agricultural tax system that annually delivered about 10 percent of the total farm produce to the Chinese Communist government, as well as a state procurement system in which the cooperatives were required to sell to the state most of their output in excess of seed, fodder, and consumption requirements. Although abolishing land payment, the advanced cooperatives provided social insurance by using public welfare funds to guarantee the livelihood of widowers, widows, orphans, aged people who had no relatives and those who were disabled, and to subsidize members who had to support big families that lacked farmhands.[11]

Thus in an advanced cooperative both production and distribution were under the control of the cooperative—the fundamental goal of collectivization.

RATE OF COOPERATIVIZATION

The rate of agricultural cooperativization had been a matter of dispute within the party since the early years of the Communist regime. The central issue was timing. One point of view, later associated with

Liu Shao-ch'i and Teng Tzu-hui, who was in charge of agrarian affairs in the party's Central Committee, advocated a gradual process. It opposed rapid cooperativization on two main grounds. First, under their new, favorable circumstances, individual farmers would increase production and, when 70 to 80 percent of them had reached the income status of "rich farmer," they would be ready for cooperativization. Second, mechanization was an essential prerequisite for cooperativization, both as an incentive for farmers to join collectives and to provide the only economic advantage considered realizable from large collective farms.

Mao Tse-tung held the opposing view that once land reform had achieved its political purpose, the farm economy should be immediately replaced by socialized agriculture. The "socialist transformation" should be completed in the shortest possible time. Reasons cited for this view were that immediate cooperativization was easier because land reform had eliminated the landed class and had reduced the influence of the old "rich peasants"; the new rich and middle farmers had not yet appeared; therefore, the sooner cooperativization was carried out, the less resistance there would be to it. Mao's statement of 1953—"With regard to the battle position in the rural areas, if socialism does not take it, capitalism will definitely occupy it"—was based, according to one explanation, on his opinion that a backward peasant economy of small farms could hardly survive; the small farms would eventually be transformed into either large socialist farms or large capitalist farms.

The opposing views clashed directly in late 1953. At a top-level meeting of the Central Committee of the CCP held on December 16, 1953, Mao threatened with possible purge those officials who were responsible for ordering the dissolution of cooperatives that had appeared in considerable numbers earlier that year, after the public announcement of a party resolution originally adopted in 1951.[12] That resolution had emphasized mutual-aid teams for the time being and had discouraged the premature organization of cooperatives. The outcome of the December 16 meeting was a new resolution, by which it was decided to set up primary APCs throughout the countryside, a compromise increasing the scheduled pace of cooperativization but reiterating gradualism.[13] It stressed voluntary participation and persuasion in its implementation through the establishment of one or more model cooperatives in each locality to demonstrate the superiority of cooperative farming.

Events in mid-1955 resulted in the escalation of the socialist transformation of agriculture. The First Five-Year Plan, published on July 5, 1955, revealed that by 1957, one-third of the farm households would be organized into primary APCs, that completion of the whole process of cooperativization was visualized in 15 years, and that cooperativization was seen as a means of raising farm production without state investment. The plan embodied the 1953 compromises. In May and June 1955, however, some 200,000 "unstable" cooperatives (a term applied by Liu Shao-ch'i, Teng Tzu-hui, and others who thought that the movement had gone faster than scheduled) were dissolved by a government directive of "drastic compression."

On July 31, 1955, just after he had returned from an extensive tour of the provinces, Mao Tse-tung, as chairman of the CCP, called for increasing the pace of agricultural cooperativization in a speech, entitled "On the Question of Agricultural Cooperation," to the secretaries of provincial, municipal, and autonomous region party committees, bypassing the regular decision-making organs (the Politburo and the Central Committee).[14] This was the first complete expression of his theory on collective farming to be made public. In it he cited all his reasons for a "socialist upsurge in China's countryside" and repudiated the so-called empiricism that was based on the analogy of the experience in the Soviet Union. Mao contended that such a method underestimated the socialist tendency of Chinese farmers and the ability of the party to lead them. He held, as before, that cooperativization must precede mechanization, not vice versa. He demanded that by 1958, 50 percent of China's farmers should have joined primary APCs and that the rural areas should be basically cooperativized by 1960. He further demanded that all APCs should become of the advanced type (collective farms) during the Third Five-Year Plan period (1963-67).

Mao's tactics resulted in the adoption, at an enlarged meeting of the Sixth Plenary Session of the Seventh National Congress of the CCP (October 11, 1955) of the "Resolution on the Question of Agricultural Cooperation" along the lines of his July 31, 1955, speech. The immediate change brought about by the decision was moderate increase in goals for early cooperativization: primary cooperatives would be formed in a majority of localities by the spring of 1958, with 70 to 80 percent of all farmers to be in such cooperatives by 1960.

As a result of Mao Tse-tung"s report " On the Question Of Agricultural Cooperation" and the adoption of the "Resolution on the Question of Agricultural Cooperation" by the Central Committee of the CCP, which was based on Mao's report, a "high tide of socialist cooperation" took place in the latter part of 1955. During this period 1.25 million new primary cooperatives were established with 70 million households, or more than 60 percent of the country's households (110 million). The total number of APCs thus increased to 1.9 million at the end of 1955.[15] These APCs were composed of 75 million farm households, about 63 percent of the country's total. Of these, 17,000 were of the advanced type and comprised more than 4.7 million farm households.[16]

On January 31, 1956, it was reported that 92.81 million farm households, 78 percent of the country's total, had participated in APCs; and 24.59 million or 26.5 percent of these households, were

TABLE 2.1

Number of Farm Households Participating in Agricultural Cooperativization, 1950-56

Year	Number of Farm Households in Mutual-Aid Teams	Number of Farm Households in APCs			Number of Farm Households in Mutual-Aid Teams and APCs
		Primary	Advanced	Total	
1950	11,313,000	187	32	219	11,313,000
1951	21,000,000	1,588	30	1,618	21,002,000
1952	45,364,000	57,000	2,000	59,000	45,423,000
1953	45,637,000	273,000	2,000	275,000	45,912,000
1954	68,478,000	2,285,000	12,000	2,297,000	70,775,000
1955	60,389,000	16,881,000	40,000	16,921,000	77,310,000
1956	—	10,407,000	107,422,000	117,829,000	117,829,000

Source: State Statistical Bureau, Ten Great Years (Peking, 1960), pp. 34-35.

TABLE 2.2

Percentage of Farm Households Participating in Agricultural Cooperativization, 1950-56

Year	Farm Households in Mutual-Aid Teams	Farm Households in Agricultural Cooperatives			Farm Households in Mutual-Aid Teams and Cooperatives
		Primary	Advanced	Total	
1950	10.7	—	—	—	10.7
1951	19.2	—	—	—	19.2
1952	39.9	0.1	—	0.1	40.0
1953	39.3	0.2	—	0.2	39.5
1954	58.3	2.0	—	2.0	60.3
1955	50.7	14.2	—	14.2	64.9
1956	—	8.5	87.8	96.3	96.3

Source: State Statistical Bureau, Ten Great Years (Peking, 1960), pp. 34-35.

members of advanced cooperatives. By mid-April 1956, 90 percent of all farm households were reported to belong to cooperatives, and 56 percent of them to have joined the advanced cooperatives. Hopei, Shansi, Kirin, Heilungkiang, Peking, and Shanghai had completed socialist cooperativization.[17]

Encouraged by the progress of implementation of his program, Mao Tse-tung called a Supreme State Conference on January 25, 1956, at which he announced his new plan to complete primary agricultural cooperativization by the end of 1956, and to complete it basically in its advanced form by 1958. This new plan shortened the period of his original plan, as stated on July 31, 1955, by nearly two years. Following the adoption of the "Model Regulations for Advanced Agricultural Producers' Cooperatives" by the Central Committee of the CCP on June 30, 1956, the pace of cooperativization was further accelerated. By the end of 1956, 120 million farm households, 96 percent of all the farm households in China, had joined APCs. More than 100 million of them, or 88 percent, joined the advanced type. The socialist transformation of agriculture was considered basically completed.[18] (See Tables 2.1 and 2.2.)

NOTES

1. Mao Tse-tung Hsuan Chi [Selected Works of Mao Tse-tung] (Peking: People's Press, 1969), p. 885.
2. The Common Program (Peking: Foreign Languages Press, 1952), Article 34, p. 14.
3. Agriculture in New China (Peking: Foreign Languages Press, 1953), p. 14.
4. Planning Commission, Report of the Indian Delegation to China on Agrarian Co-operatives (New Delhi, 1957), p. 34.
5. Agricultural Policy Department, Ministry of Agriculture, Central People's Government, "Certain Views Concerning the Convention of the Mutual-Aid Teams," in its Reference Materials on the Mutual-Aid Teams for Agricultural Production, I (Peking: Ministry of Agriculture, Central People's Government, 1952), pp. 1-4.
6. Yao Kung-chen, Questions Concerning the Economic Reconstruction in Rural China (enl. and rev. ed.; Shanghai: Yung-hsiang Press, 1953), p. 4.
7. Mao Tse-tung, On the Question of Agricultural Cooperation (English ed.; Peking: Foreign Languages Press, 1966), p. 3.
8. Ibid., p. 34.
9. "Communique' on the Results of the Implementation of the 1955 National Economic Plan," Hsin-hua Pan-yueh-k'an, Peking, No. 13, 1956, pp. 39-42.

10. Mao Tse-tung, On the Question of Agricultural Cooperation (2nd rev. ed.; Peking: Foreign Languages Press, 1966), p. 35.

11. Frederick W. Crook and Linda A. Bernstein, Agriculture in the United Stated and the People's Republic of China. 1967-71. Foreign Agricultural Economic Report No. 94 (Washington, D.C.: Economic Research Service, U.S. Department of Agriculture, 1974), p. 5.

12. "Decision on Mutual Aid and Cooperation in Agricultural Production," adopted by the Central Committee of the CCP in December 1951. The first draft was issued on December 15, 1951, to local party organizations to be tried in various places, but was not published in the press as a formal party decision until March 1953. See Mao Tse-tung, On the Question of Agricultural Cooperation, p. 4.

13. "Resolution on the Development of Agricultural Producers' Cooperatives," adopted by the Central Committee of the CCP on December 16, 1953; published as Cooperative Farming in China (Peking: Foreign Languages Press, 1954).

14. JMJP, October 17, 1955, p. 1.

15. Mao Tse-tung's preface to Socialist Upsurge in China's Countryside, compiled by Central Committee of the CCP (Peking: Foreign Lanugages Press, 1957), p. 8; and State Statistical Bureau, Ten Great Years (Peking, 1960), pp. 26-28.

16. State Statistical Bureau, Communiqué on the Results of the Implementation of the 1955 National Economic Plan (Peking, 1956); JMJP, June 15,1956, pp. 1, 3; Hsin-hua Pan-yueh-k'an, Peking, No. 13, 1956, pp. 39-42.

17. Ch'en Po-ta, Chung-kuo Nung-yeh ti Sheh-hui Chu-i Kai-tsao [Socialist Transformation of China's Agriculture] (Peking: People's Press, 1956), p. 2.

18. Socialist Upsurge in China's Countryside, p. 8; State Statistical Bureau, loc. cit.

CHAPTER

3

COMMUNALIZATION AND ITS MODIFICATION

FROM COOPERATIVIZATION TO COMMUNALIZATION, 1957-58

Perhaps encouraged by the generally good harvests of 1957, the Central Committee of the CCP decided "to accelerate socialist construction, to complete the building of socialism ahead of time, and to carry out the gradual transition to communism."[1] As a first step to attain these goals, the committee, at a meeting held at Chengtu in March 1958, adopted a proposal submitted by Mao Tse-tung to merge small cooperatives into larger ones; and "Directive on the Questions of Merging Cooperatives" was promulgated shortly afterward. Mao's intention to merge cooperatives had actually been revealed earlier. In his comments on one of the articles printed in The Socialist Upsurge in China's Countryside, he said:

> Most of our present semi-socialist cooperatives have only twenty or thirty families in them, because cooperatives of that size are easy to run and they give the administrative staff and the members an opportunity to gain experience quickly. But small cooperatives have fewer members, less land and not much money. They cannot operate on a large scale or use machinery. The development of their forces of production is still hampered. They should not stay in this position too long, but should go on to combine with other cooperatives.[2]

From the winter of 1957 to the summer of 1958, extensive improvements were made on farmland, centered on building water conservation projects. Much also was done to develop agriculture-oriented industry, transport and communications, commerce, culture, education, and health. The advanced cooperatives, organized chiefly for farming, became increasingly inadequate for performing such multiple functions. In many places the smaller cooperatives amalgamated themselves into big ones. During the first half of 1958, about 789,000 APCs were merged into 741,000.[3] Since such a cooperative often embraced the households of an entire township, its managing group was combined with the township government, resulting in a unit of both political and economic organization. This was the prototype of the people's commune.[4]

An an experiment the Weihsing (Sputnik) People's Commune in Suiping hsien, Honan Province, was established in April 1958 by the merger of 27 APCs covering four townships. It had 9,300 households and 43,000 people.[5] Early in August 1958, Mao Tse-tung visited a number of areas in Honan, Shantung, Anhwei, Peking, and Tientsin to inspect local agricultural conditions; and on August 29, 1958, he issued the statement "People's communes are good." On the same date "Resolution on the Establishment of People's Communes in the Rural Areas" was adopted by the Central Committee of the CCP, officially inaugurating the commune system. It was announced on October 1, 1958, that the rural areas throughout the country had been basically communalized by merging 740,000 APCs into 26,000 people's communes; and about 120 million farm households, which constituted more than 90 percent of China's rural population, had joined the communes.[6]

CHARACTERISTICS OF THE PEOPLE'S COMMUNE

Compared with the advanced APCs, the communes were bigger and more public. With more land, more people, and more funds, they were said to be able to use manpower and other means of production more rationally and to carry out a diversified economy more efficiently. The communes were considered more public in the sense that they had a higher degree of common ownership than the advanced cooperatives had and, therefore, could strengthen the rural collective economy and speed up the building of socialism in the countryside. They could do a better job of improving land on a large scale, experimenting with scientific farming, and fighting natural disasters. Their far greater economic strength could make possible faster progress in water conservation and other major aspects of the technical transformation of agriculture.[7]

More specifically, at its beginning the commune had the following characteristics, according to the People's Daily:[8]

1. The commune was an organization not solely engaged in agricultural production, nor did it deal only with agriculture, forestry, animal husbandry, fishery, and side occupations. It also built industry, aiming at the gradual elimination of the difference between urban and rural areas and between industry and agriculture. Besides production, the commune engaged in trade (exchange) and credit (banking). In addition to economic activities, it undertook the tasks of culture and education. It also carried out the system of citizen soldiery. Young men and retired servicemen were organized into the militia. Thus the commune took charge of political, economic, education, cultural and military affairs at the same time.

2. Having the same area as a township, the commune was much bigger than the APC. Early in 1958 each of the more than 740,000 APCs had an average of 170 households, 300 laborers, and 2,000 mou (333 acres) of land. By contrast each of the 26,000 communes in the latter part of 1958 had an average of nearly 5,000 households, 10,000 laborers, and 64,000 mou (10,667 acres) of land. A commune was thus about 30 times as large as the average advanced APC. The pooling and unified control of manpower and other resources were believed to enable the commune to facilitate agricultural capital construction, develop a diversified economy, and speed up the modernization of agriculture.

3. The commune and the township were merged. Deputies to the township people's congress were concurrently representatives to the congress of the commune; members of the township people's council were concurrently members of the administrative committee of the commune; the head and deputy heads of the township were concurrently the head and deputy heads of the commune; and the departments under the administrative committee of the commune were concurrently the departments under the township people's council. Thus the commune was the basic political organ of the state in the PRC's countryside and in this respect, too, it differed from the APC, which was primarily an economic organization.

4. The commune went a step further than the APC toward the development of common ownershp. All land reserved for private use, private house sites, livestock, and trees owned by individuals were gradually turned over to common ownership. The small number of domestic animals and poultry temporarily kept by individuals were also to be gradually turned over to common ownership.

5. In the distribution of income and remuneration of labor, the commune would gradually, with the development of production, substitute the wage system for the labor-day bonus system. The grain supply system was to be put into practice when grain production became highly developed and when the members of the commune unanimously agreed to it. This system was believed to provide additional guarantees for the income and livelihood of commune members, facilitate expanded production, and be more beneficial to the develop-

ment of industry, transport and communications, education, science, culture, and welfare services.

6. Welfare services, including organizations for facilitating production and liberating women's labor power, such as public dining halls, community canteens, nurseries, and sewing teams, were developed in the commune. Medical and public health organizations; welfare institutions for orphans, widows, old people, and the disabled; and new housing quarters were established.

7. Allowing for local or regional variations, the organization of the commune generally consisted of three levels: the commune as a whole, the production brigade, and the production team. Most of the communes had production brigades and production teams, although some of the smaller communes had only production teams. At the end of 1960, there were more than 24,000 communes, 500,000 production brigades, and 3 million production teams.

The commune's highest governing body was its congress, which decided on all matters important to the communes as a whole. Members of the congress included representatives of production brigades and various population sectors: the women, youth, old people, and workers in various trades, industries, and professions. The congress elected an administrative committtee to carry on functions of the commune as a whole and to supervise the work of production brigades and production teams. Under the administrative committee were the departments in charge of specific functions, such as agriculture, water conservation, animal husbandry, rural industry, communications, commerce, finance, culture and education, scientific research, and the militia.

Thus the commune was designed to be "a transitional form of organization from collective ownership to state ownership (ownership by the whole people)," or "a form of organization for building up socialism and gradually passing on towards communism." It was intended to be the testing ground for a totally collective life, preparing the way for the development from the socialist system to the Communist system.[9]

MODIFICATION OF THE COMMUNE SYSTEM, 1960-65

The decision to communalize agriculture was made in haste, without advance planning; and the central authorities were unable to provide detailed operational instructions to cadres about the form, structure, size, and functions of communes. They had to confine their aims to constituting units that were allowed a large measure of autonomy in the choice and appointment of their organization. Consequently cadres were compelled to improvise and make their own decisions. Disorder and confusion arose from divergent practices in different communes. Meanwhile, some of the peasants jibed at the excessive collectivization, which was extended to every sphere of activity.

It was the original intention of the central authorities to maintain, as much as possible, the collective ownership of means of production and the unified distribution of income and products under the commune system. It did not take long, however, for many high officials to realize that such centralization of power would not help agricultural production, and that the original communalization plan and the major functions of the three levels of the commune had to be readjusted.

Disorganization of labor, caused by the sudden change in 1958 from the APCs to the communes, had resulted in a general breakdown of production and distribution systems. Shortage of skilled reapers in some rural areas had delayed harvests. In some areas, for instance, the peasants were still picking cotton in late November 1958, when frosts had damaged the bolls and had made them unfit for processing.

Payment to the peasants according to the amount of their work was almost impossible, not only because of the centralized deployment of labor but also because of the lack of an efficient accounting system. Many party cadres during the Leap Forward thought that such bagatelles as accounting and bookkeeping could be ignored. The neglect of financial management produced not only disorder in receipt and disbursement, but also extravagance and waste. Agricultural production, as well as capital accumulation and distribution in the communes and the livelihood of commune members, was badly affected.

The agricultural situation deteriorated in 1959-61. The Communist government blamed natural calamities—particularly drought, floods, and pests—as solely responsible for the three consecutive years of bad harvests; but it was apparent that the crop failure was also due to the mismanagement of agricultural production and to the peasants' lack of initiative to work under the commune system.

Beginning in the latter part of 1959, it was necessary for the Communist authorities to readjust certain functions of the three levels of the commune. The modifications were motivated by the fact that the complete regimentation of peasants had not increased production, and were based on a resolution adopted at the Seventh Plenary Session of the CCP's Eighth National Congress, held at Shanghai in April 1959. Under the liberalized policy, the production brigade gradually became the basic unit of accounting and auditing. It also became the unit for the unified distribution of income and of agricultural and sideline products of its production teams. It owned most of the means of production—including land, draft animals, and farm implements—within its borders. It ran most of the commune-owned industries. It assigned agricultural production work to its production teams. The production brigade also enjoyed considerable latitude in such matters as the allocation of its funds for investment in capital construction or other needs.

Meanwhile, the commune became a unit that owned only a small part of the means of production and agricultural and sideline products, and was mainly concerned with the direct operation of industrial establishments that were collectively owned by all members of the commune, such as chemical fertilizer plants and farm-implement repair plants. The production team served only to attend to the actual agricultural production work. It used land and, in most cases, draft animals and farm implements that had been assigned to it by the production brigade. It also was mainly responsible for the operation of certain welfare activities, such as nurseries, kindergartens, canteens, and public dining halls.

An emergency directive of 12 articles on rural work, issued by the Central Committee of the CCP on November 3, 1960, granted further concesssions to the peasants.[10] In order to permit the peasants to exercise more initiative, to suffer less interference, and to receive a larger share of income and output, the original organization of the communes, as it existed in 1958-60, was further modified in the "Draft Regulations Governing the Work of the Rural People's Communes," adopted by the Central Committee of the CCP on May 12, 1961, and revised in September 1962. The new regulations, consisting of 60 articles that expanded and restated more specifically the contents of the November 3, 1960, directive, provided for, among other things, (1) placing the decisions on production in the hands of the production team and curtailing the powers of the commune and the production brigade, which had led to the abuse of labor mobility and frequent changes in production plans; (2) again permitting commune members to till small private plots for their own account (the resulting agricultural products were placed at their disposal and were not distributed as rations or used for state collection and purchase); (3) permitting commune members to raise pigs, sheep, and domestic fowls, to take up family handicrafts and sideline occupations, to grow vegetables and fruit trees around their houses, and to keep the income from these activities; (4) permitting commune members to sell the things they produced and to buy the things they needed at rural fairs; (5) basing compensation more closely on the amount of work performed (implementing the principle of distribution according to labor); and (6) increasing the consumption of commune members in proportion to the total of production investment, administrative expenses, communal accumulation, and agricultural tax, and their monetary compensation in proportion to rations in kind.[11]

UNIVERSITY
OF BRISTOL
LIBRARY

The May 1961 regulations, however, still reflected a certain measure of hesitancy or reluctance on the part of the Communist authorities to relinquish too much control. For example, most power over output distribution was still vested in the production brigade; and the production brigade and, to a lesser extent, the commune could still decisively influence the scheme of production and distribution. Thus, on August 29, 1961, the People's Daily wrote enthusiastically on the production brigade as the independent accounting unit, in contrast with the production team, which was the basic unit in organizing production.[12]

Disappointment over the small increase in agricultural output in 1961, together with popular discontent, made it necessary for the government to downgrade the production brigade and to upgrade the production team in the latter part of that year. Distribution of income and ownership of the land and most of the equipment and livestock it used were vested in the hands of the production team. In the New Year editorial on January 1, 1962, the People's Daily took the line that agricultural recovery was dependent upon the "production activism" of the masses and that this was to be accomplished through implementation of the party's policy of using the production team as the accounting unit. In another editorial[13] the same paper spoke of the overexpansion of the production brigade, which led to egalitarianism in distributing output, to the detriment of incentive.

In compliance with the new policy, the May 1961 regulations were revised at the Tenth Session of the Eighth Central Committee of the CCP, which met in September 1962. By 1963, with the exception of a few areas, the production team had become the basic unit for both accounting and production. It was accorded the right to keep its own accounts, to map out production plans, to prepare its budget, and to distribute income to its members. The major functions of the production brigade were reduced to organizing cooperation among the production teams and running enterprises jointly owned by them. The commune was assigned the responsibilities of carrying out large agricultural projects, and activities of a commune-wide or intercommune nature. Moreover, the average size of most of the communes was reduced to approximately that of the former advanced APCs and their number was increased from 24,000 to 74,000 because large communes under centralized management had proved inefficient in carrying out agricultural production. Some 700,000 production brigades were formed from the original 500,000, and 5 million production teams from the original 3 million. In effect, therefore, each old large commune was divided into three, each two production brigades turned into three, and each production team made smaller (only 10 to 20 households).

While adopting this more liberal policy, however, the government from time to time called for the consolidation of the collective agricultural economy. Although there was no open reversal of the policy of relaxation and decentralized control, it would seem that any substantial recovery of agricultural production might engender fear by the authorities that the peasants, when better off, might resist the return to collectivism. For example, although the decision reached by the Tenth Session of the Eighth Central Committee of the CCP in September 1962 continued the policy of emphasizing the production team, it also called for further consolidation of the collective economy and the promotion of the peasants' enthusiasm for collective production.[14] The government repeatedly made the point in its propaganda that communes, with their abundant capital, labor, and other resources, are more effective in fighting natural calamities and in carrying out large-scale projects than the APCs were.

NOTES

1. "Resolution on the Establishment of the People's Communes in the Rural Areas," People's Communes in China (Peking: Foreign Languages Press, 1958), pp. 1-8.

2. Municipal Committee of Hsinhailien, Kiangsu Province, Chinese Communist Party, "The Superiority of Large Cooperatives," in The Socialist Upsurge in China's Countryside, compiled by the General Office of the Central Committee of the Chinese Communist Party (English ed.; Peking: Foreign Languages Press, 1957), pp. 460-76.

3. Editorial, JMJP, September 25, 1958, p. 1.

4. Some Basic Facts About China, supplement to China Reconstructs, no. 1, Peking, January 1974, pp. 4-5.

5. JMJP editor's note to "Tentative Regulations (draft) of the Weihsing People's Commune, August 7, 1958," translated in People's Communes in China, pp. 61-77.

6. State Statistical Bureau, Ten Great Years (Peking, 1960), p. 43.

7. Some Basic Facts About China, loc. cit.

8. Editorial, JMJP, September 4, 1958, p. 1.

9. For the development of a commune in Hsinhsiang hsien, Honan Province, since its establishment in 1958, see Chu Li and Tien Chieh-yun, Inside a People's Commune (Peking: Foreign Languages Press, 1974).

10. The directive was monitored for domestic use only in a broadcast from the Kiangsi People's Broadcasting Station on November 3, 1960, and was mentioned in a JMJP editorial, "Sum up the Experiences of Communes and Strengthen the Development of Communes," November 20, 1960, p. 1.

11. JMJP, May 13, 1961, p. 1.

12. "Each Should Assume Its Responsibilities and Each Should Develop Its Ability," editorial, JMJP, August 29, 1961, p. 1.

13. "The Glory of Collectivism," editorial, JMJP, January 4, 1962, p. 1.

14. JMJP, September 29, 1962, p. 1.

CHAPTER

4

TWO MOVEMENTS

THE RURAL SOCIALIST EDUCATION MOVEMENT, 1963-65

The adoption of a series of moderate measures in 1960-62 revived rural economy somewhat; but it also gave rise to speculation and corruption on the part of many basic-echelon rural cadres, which not only affected the soundness of the collective economy but also adulterated the "purity of the cadre forces." The influence of free economy was said to have gradually infiltrated into the economic activities of the people, and the cadres' ideology was said to have been corroded by capitalistic thinking. Beginning in early 1963 the Communist authorities were compelled to launch the Four Cleanups Movement, also known as the Rural Socialist Education Movement, which at first referred to the cleaning up of work points, accounts, properties, and warehouses through propaganda and education of the peasant masses and the basic-level cadres, but later (1965-66) was broadened to include politics, ideology, organizations, and the economy in general.[1]

Although most of the CCP leaders agreed on the necessity of such a movement, there was sharp contradiction on policies, strategies, and methods of approach between Mao Tse-tung and Liu Shao-chi'i. Some of their conflict was revealed or reflected in several documents adopted by the Central Committee of the CCP during 1963-66 that were not made public until the Cultural Revolution. In September 1962, when the nation's economy began to show signs of recovery, Mao Tse-tung reminded the cadres attending the Tenth Plenary Session of the Eighth Central Committee of the CCP of the importance of class struggle. In accordance with his recommendation, the committee adopted "Draft Regulations on Some Problems Existing in Current Rural Work" on May 20, 1963. The document, known during the Cultural Revolution as the "Former Ten Points," reportedly was prepared by Mao. It brought to light serious and sharp class contradictions existing in rural communities. The document pointed out that education of the rural masses, the strengthening of unity between the party and the masses and between the cadres and the masses, and the consolidation of the collective economy and the proletarian dictatorship would depend upon the Four Cleanups Movement of the poor and lower-middle peasants in relation to the cadres and their struggle against the class enemies (landlords, rich peasants, counterrevolutionaries, and local undesirable elements, such as thieves and criminals). The May 20, 1963, document laid down two major guidelines for the rural socialist education movement: (1) the establishment of poor and lower-middle peasants' associations at the commune, production brigade, and production team levels, as a subsidiary of the party to supervise the work of the basic-level cadres, and (2) the systematization of cadre participation in farm labor to guarantee sound direction of production and to prevent the alienation of the cadres from the masses.[2]

The campaign began to take shape in the late spring of 1963. That September the CCP Central Committee, on the basis of the four-month experience, adopted another important document, "Some Concrete Policy Formulations Concerning the Rural Socialist Education Movement (draft)." This document, known as the "Latter Ten Points," reported was written by Liu Shao-ch'i. It dealt with the "ten problems supplementing the contents of the socialist education movement which had been carried on in various rural areas since the Tenth Plenary Session of the Eighth Central Committee." "The ten problems" apparently referred to the "Former Ten Points" drafted by Mao Tse-tung. The document marked the beginning of serious persecution and declared that "the socialist education movement has its long-range and far-reaching significance in repulsing the frantic offensive of the once loudly aggressive imperialistic and feudalistic forces, in consolidating the position of rural socialist and proletarian dictatorship, in destroying the social basis from which revisionism stems, in consolidating collective economy, and in developing agricultural production."[3]

In an attempt to strengthen his viewpoint, in November 1963 Liu Shao-ch'i sent his wife, Wang Kuang-mei, to the Taoyuan Production Brigade of the Luwangchuan Commune in Funing hsien, Hopei Province, to investigate the situation there. In approximately five months she completed the investigation, during which she was reported to have discovered serious cadre corruption and other problems. Her findings, often referred to as the "Taoyuan experiences," were incorporated in Liu Shao-ch'i's "Revised Latter Ten Points," promulgated in September 1964.[4]

Meanwhile, Mao Tse-tung issued "Directive Listing Six Criteria for Evaluating the Success or Failure of the Rural Socialist Education Movement" in June 1964. This was followed by the adoption of the "Organization Rules of Poor and Lower-Middle Peasants' Associations" which supplemented Mao's directive. Beginning in late 1964, poor and lower-middle peasants' associations were organized by the Communist party authorities of many provinces and municipalities, with emphasis on class struggle.[5]

At the beginning of 1965, the Maoists claimed that the Rural Socialist Education Movement had penetrated to every front in the cities as well as in the rural areas, and that in the course of the campaign, all the basic cadres had been affected, many of them being dismissed or punished. On January 14, 1965, another policy document, "Some Problems Currently Arising in the Course of the Rural Socialist Education Movement," was adopted by the National Working Conference convened by the Central Political Bureau. This 23-point directive, reportedly worked out by Mao Tse-tung, defined the exact nature of the campaign, with emphasis on capitalists holding power inside the party. According to the Maoists, the document was designed to repudiate and correct Liu Shao-ch'i's erroneous line embodied in the "Latter Ten Points," which was Left in form but Right in essence and which had sabotaged the Rural Socialist Education Movement. It also was reported that in the document, which was never published, Mao for the first time specifically stated: "The main target of the present movement is those Party persons in power taking the capitalist road."[6]

Toward the end of 1965, the Maoists launched the Great Proletarian Cultural Revolution with an attack against the leading cadres in the Peking Municipal Party Committee. The attack soon turned into a stormy campaign that completely overshadowed the Rural Socialist Education Movement.

THE "LEARN FROM TACHAI" MOVEMENT, 1964–

Since early 1964 Tachai production brigade of Tachai Commune in Hsiyang hsien, Shansi Province, has become a pacesetter for the country's agriculture by virtue of the hard work and the self-reliant spirit that it applied to turning its loess hill into vast staircases of terraced fields. Tachai is situated in the seriously eroded Taihang Mountains. In 1971 it had 83 households with a population of 430. The area is cut by seven ravines and dozens of gullies. The ravines and gullies are dry most of the time, but in the rainy season raging torrents used to bite into the hillsides and damage farmland on the slopes.[7]

Through sheer human effort Tachai reportedly has achieved spectacular agricultural success. In ten years of hard battle, the peasants of Tachai put in 80,000 labor days and finally succeeded in turning the barren slopes into terraced fields. As a result grain production, family income, property values, and living standards were said to have increased considerably.[8] In 1960 the Tachai production brigade was chosen by the Shansi Provincial Party Committee as an example to be imitated by the whole province. At the Third National Congress held in December 1964, Chou En-lai called upon the whole country to learn from Tachai; and shortly afterward the slogan of "In agriculture, learn from Tachai" was put forth by Mao Tse-tung.[9]

In the following two years a mass movement in response to this call took shape. However, it was not until 1967, when Liu Shao-ch'i's influence in opposition to the movement began to wane, that the slogan was circulated more frequently. For the next few years a general mobilization for the movement was launched in the countryside. Today Tachai has become a showplace of China's agriculture; and many visitors, including rural cadres, poor and lower-middle peasants, and foreign guests, have come to learn from its experience.

With the progress of the "Learn from Tachai" movement, a peasant leader in that production brigade, Ch'en Yung-kuei, has become a national hero. Ch'en began his political career in the early 1960s, when he served as party branch secretary of both Hsiyang hsien and the Tachai production brigade. In 1969 he was elected a member of the CCP Central Committee.[10] He became one of the secretaries of the CCP Committee for Shansi Province in 1971.[11] Two years later Ch'en was elected a member of the CCP Central Political Bureau at the Tenth National Party Congress. He was named a member of the presidium of the First Session of the Fourth National People's Congress held on January 13-17, 1975, and was appointed one of nine vice-premiers of the State Council at that session.[12]

Since early 1971 Ch'en has been sent to several provinces to publicize his doctrines and the accomplishments of Hsiyang hsien and the Tachai production brigade; he also has visited Mexico, Romania, and Canada at the invitation of the governments of those countries.[13] His articles have appeared in _Hung Ch'i_ and other leading magazines, and his book, _Talking About Scientific Cultivation of the Land_, was published by the Agriculture Press of Peking in 1973.[14]

According to Ch'en the "Learn from Tachai" movement means politics in command, self-reliance, and bitter struggle. The Tachai spirit requires that one should fear neither hardship nor death, and should work for the revolution, not for remuneration. Peasants should be paid in proportion to political loyalty under the piao-ping kung-feng (standard-bearer work points) stystem, which Tachai introduced and in which the thought of Mao Tse-tung is more important than remuneration.[15]

For the past decade or so, Ch'en's views generally have reflected the changing economic policy for China's countryside. In the years of the Cultural Revolution (1966-69), when the agrarian policy was extremely restrictive, the Tachai system meant remuneration of the peasants primarily in proportion to their political loyalty. When an economically more rational policy was introduced early in 1971, the message of Tachai changed accordingly. "Directive on the Question of Distribution in the Rural People's Communes," issued by the Central Committee of the CCP on December 26, 1971, for internal use only, stipulated that the Tachai-style labor management and work-point system should not be enforced compulsorily: "In studying the labor administration of the Tachai-type commune, attention must be paid to the actual conditions. The masses should be consulted and not be forced with preconceived methods. The masses must be taught to seriously assimilate the model experiences of the local areas. Those extraneous and unreasonable things which restrict the initiative of the masses must be done away with. Methods should be employed which are popular with the masses and which can be easily put into practice. These methods should be constantly improved upon."[16]

Regardless of the changing economic policies, the fundamental lesson of the "Learn from Tachai" movement remained the same politics in command, self-reliance, and bitter struggle. The movement went into full swing in the second half of 1970. Under political pressure many provinces selected one or several production brigades as their own Tachai. Toward the end of 1970, however, the People's Daily complained that in many places a local Tachai had been designated but agricultural production had not increased, "because the class struggle had not been carried out properly," that is, not enough pressure had been put upon the cadres and the peasants.[17] News came from many provinces about resistance to the introduction of the Tachai system,[18] but the pressure for its implementation continued.

Using Tachai as a pattern, communes throughout the country have recently carried out "farmland capital construction work" on an unprecedented scale. Such work was usually done after the autumn harvest, in preparation for farming the following spring. Some 100 million people took part in various projects in each winter-spring period during 1971-75, bringing an average of 1.6 million more hectares of land under irrigation each year.[19] In the three north-eastern provinces of Heilungkiang, Kirin, and Liaoning, the number of participants during the winter of 1973-74 was more than 4 million, and Kwangtung Province it was over 10 million. In the fourth quarter of 1973, communes throughout the country built or improved irrigation facilities for some 1.3 million hectares of farmland. At the same time 2.6 million hectares of farmland were leveled, 12.6 million hectares were deep-plowed, and 460,000 hectares were terraced.

The purpose of farmland capital construction work is to transform land conditions, with a view to increasing agricultural production. The work usually is related to water and soil conservation. It is not only aimed at achieving higher yields in the forthcoming harvest but also is closely linked with the long-term plans to lay foundations for a large-scale, modernized agriculture. In the Yangtze and Pearl river deltas and many areas on the plains, farmland has been laid out in rectangular plots flanked by straight ditches and roads suitable for mechanized farming. In many hilly regions small plots have been combined into larger ones suitable for the operation of small mechanized tractors and hand tractors.[20]

For summing up experience gained in the Tachai movement and in farm mechanization, a national conference, considered "extremely important" by the authorities, was held in Hsiyang hsien (the closing sessions were held in Peking) on September 15-October 20, 1975, under the auspices of the State Council. Among the 3,700 people attending the conference were the vice-premiers Teng Hsiao-p'ing, Hua Kuo-feng, and Ch'en Yung-kuei, and Chiang Ch'ing, member of the Politbureau of the Central Committee of the CCP. The conference reviewed the accomplishments in the 11 years since Mao tse-tung's call for learning from Tachai in agriculture in 1964; and discussed how to make further efforts in launching the movement and how to fulfill the task of farm mechanization in the main by 1980, the final year of the Fifth Five-Year Plan.

In his opening speech Ch'en Yung-kuei said that the "Learn from Tachai" movement was "developing unevenly from one province to another and from one hsien to another. The gap in agricultural production remained very wide." He urged that there be faster agricultural development.[21]

The conference announced an ambitious program of agricultural development for the Fifth Five-Year Plan (1976-80) and put forward a call for the whole party to mobilize, to go all-out to develop agriculture, and to build up Tachai hsien throughout the country.[22] The conference demanded that one-third of the country's more than 2,000 hsien should be transformed into Tachai-type hsien by 1980, in order to further consolidate the dictatorship of the proletariat and the socialist system, accelerate China's socialist construction, and strengthen its power to smash enemy attacks in any form.[23] At least 100 hsien should become Tachai-like every year. The hsien to achieve this would be designated in each province; and each province would report on the progress twice a year, after the summer harvest and at the end of the year.[24]

In his keynote speech Vice-Premier Hua Kuo-feng (acting premier since February 7, 1976) summed up the discussions of the conference and prescribed the following six standards for a Tachai-type hsien: (1) the hsien party committee should be a leading core that firmly adheres to the party's line and policies, and is united in struggle; (2) it should establish the dominance of the poor and lower-middle peasants as a class, so as to be able to wage resolute struggles against capitalist activities and exercise effective supervision over the class enemies and remold them; (3) cadres at the hsien, commune, and brigade levels should, like those in Hsiyang, regularly participate in collective productive labor; (4) rapid progress and substantial results should be achieved in farmland capital construction, mechanization of agriculture, and scientific farming; (5) the collective economy should be steadily expanded, and production and income of the poor communes and brigades should reach or surpass the present level of the average communes and brigades in the locality; and (6) all-round development should be made in agriculture, forestry, animal husbandry, side occupations, and fishery, with considerable increases in output, major contributions to the state, and steady improvement in the living standards of the commune members.[25]

The conference set the deadline for basically achieving farm mechanization by the end of 1980—that is, 70 percent of the principal jobs in agriculture, forestry, animal husbandry, sideline occupations, and fishery would be done by machines by that time. Realizing that there would be scant national resources available to help agriculture, however, the conference reemphasized the old line of local self-reliance. The People's Daily stated: "The National Conference on Learning from Tachai in Agriculture has discussed the glorious and arduous task of how to basically mechanize farming before 1980. How can we speed up mechanization? Many delegates say that on this question there are two different schools of thought and two different lines. In one case, some people expect assistance from the state and hope to receive state loans. As a result, they have waited for many years but their natural conditions remain unchanged. Another case is marked by the effort to develop commune-owned and brigade-owned enterprises according to local resources and conditions by mainly relying on the strength of the collective economy. Once such enterprises are vigorously developed, agricultural mechanization becomes an attainable goal."[26]

The "Learn from Tachai" conference undoubtedly was one of the most important meetings held in China in recent years, since it laid down both political and agricultural guidelines. Politically there had been a renewed struggle for power between the moderate and radical factions of the CCP shortly before (and after) the death of Premier Chou En-lai on January 8, 1976, and one of the principal purposes of the conference was to strike a balance between the two factions and to work out a future path for China. Agriculturally the conference was an intensification of the "Learn from Tachai" movement launched in 1964. Rural China's productivity had increased steadily since then, but the Communist authorities felt that the pace of increase must be still higher if China is to attain its goal of becoming an industrialized power by the year 2000.

NOTES

1. For an extensive discussion on this subject, see Richard Baum and Frederick C. Teiwes, Ssu Ch'ing: The Socialist Education Movement of 1962-1966 (Berkeley: Center for Chinese Studies, University of California, 1968).

2. For an English translation of the document, see Issues and Studies, Taipei, Vol. II, No. 8, May 1966, pp. 46-60.

3. Issues and Studies, Vol. II, No. 9, June 1966, pp. 34-44; Vol. II, No. 10, 1966.

4. An English translation of the "Revised Latter Ten Points" is in Issues and Studies, Vol. I, No. 10, July 1965, pp. 1-12; and Vol. I, No. 11, August 1965, pp. 27-43.

5. Li Ming-hua, "Peasants' Associations and Class Struggle in Rural Areas in China Mainland," Issues and Studies, Vol. I, No. 7, April 1965, pp. 46-52.

6. Kuo Feng-lien, "The Tachai Road," Peking Review, No. 40, 1974, pp. 18-25.

7. Tachai: Standard Bearer in China's Agriculture (Peking: Foreign Languages Press, 1972), p. 2.

8. Kuo Feng-lien, op. cit.

9. Tao Lu-chai, "Let the Tachai Spirit Unfold Flowers and Bear Fruits Everywhere," Hung Ch'i [Red Flag], Peking, No. 11, 1965, pp. 4-14.

10. JMJP, July 3, 1967, p. 2; and April 25, 1969, p. 3.

11. Ibid., April 17, 1971, p. 1.

12. NCNA, Peking, January 20, 1975.

13. Ibid., March 27, 1975, and April 5, 1975.

14. JMJP, July 31, 1973, p. 1.

15. Ibid., editorial, September 23, 1970, p. 1.

16. See Chung-kung Yen-chiu [Studies on Chinese Communism], Taipei, Vol. VI, No. 9, 1972, Chinese text on pp. 99-104, English translation on pp. 25-28; English translation also printed in Ch'en Ting-chung, "A Study of Mao Tse-tung's Agricultural Line," Issues and Studies, Vol. X, No. 14, November 1974, pp. 77-89.

17. JMJP, December 17, 1970, p. 4.

18. Ibid., December 17, 1973, p. 2; and March 7, 1974, p. 2.

19. "Mobilize the Whole Party, Make Greater Efforts to Develop Agriculture and Strive to Build Tachai-type Counties Throughout the Country," excerpts of summing-up report by Hua Kuo-feng at the National Conference on Learning from Tachai in Agriculture, October 15, 1975, Peking Review, No. 44, October 31, 1975, pp. 7-10, 18.

20. NCNA-English, Peking, December 3, 1975.

21. "The Whole Country Is Viewed from Tachai—The National Conference on Learning from Tachai in Agriculture Discusses How to Popularize the Tachai-type Hsien as Soon as Possible," JMJP correspondent's commentary, JMJP, September 26, 1975, pp. 1-2.

22. "Build up Tachai-type Hsien Throughout the Country," JMJP, editorial, October 21, 1975, p. 1.

23. NCNA-English, Peking, November 14, 1975.

24. JMJP, October 21, 1975, pp. 2-3.

25. "Mobilize the Whole Party, Make Greater Efforts to Develop Agriculture and Strive to Build Tachai-type Counties Throughout the Country."

26. "Here Lies the Hope," JMJP correspondent, JMJP, October 17, 1975, p. 1.

CHAPTER

5

THE CULTURAL REVOLUTION AND AFTER

COMMUNES DURING THE CULTURAL REVOLUTION, 1966-69

The Great Proletarian Cultural Revolution, instigated by Mao Tse-tung in August 1966 with the aim of purging his political opponents, has been labeled by his followers as "a powerful motive force for the development of the socialist productivity in our country."[1] It is not surprising, therefore, that numerous agricultural successes have been claimed since 1966. Because of the obscurities prevailing in the overall Chinese situation during that period and since then, however, it is difficult to gauge the extent to which the Cultural Revolution has benefited or affected agricultural production. Peking has claimed that China's principal crops generally have had good or excellent harvests for the consecutive years since 1966, despite the natural calamaties in many parts of the country. The success has been attributed to "the guidance of Mao Tse-tung's thought and the soaring enthusiasm generated by the Great Proletarian Cultural Revolution."

Few, if any, of the glowing claims have thrown much light on the actual situation in agricultural production. At the beginning of the Cultural Revolution, large numbers of disgruntled peasants left their homes for the cities, ostensibly to "exchange revolutionary experiences" but actually to seek better economic conditions. Young intellectuals who previously had been sent to the countryside to take up farming also took this opportunity to return to the cities. Meanwhile, streams of workers and students in the cities were sent to the countryside to participate in farming and to "seize power."[2] The confrontation of the rural and urban groups resulted in brawls and riots. Moreover, at many communes party cadres were opposed by the peasants, so that the leadership of agricultural production was shaken. This greatly aggravated the confusion in the rural areas; and in the fall of 1966, when the situation became very serious, the peasants throughout the country were repeatedly called upon to "grasp the revolution and promote production."[3] The situation apparently had not improved when the time for spring sowing was near. In February 1967 it was necessary for the top party and government authorities to issue urgent directives to the peasants and cadres at all levels, urging a new upsurge in spring plowing at once, as well as a grasp of the Cultural Revolution.[4]

As a means to consolidate their political gains, the pro-Maoists, beginning in early 1967, established the "Revolutionary Committees" in the provinces, autonomous regions, municipalities, and lower administrative subdivisions, where power-seizure was considered completed or nearly completed. In each committee the party nucleus is the core of the leadership. These committees are the provisional Maoist-oriented origins of power, taking the place of the party committees or the governments discredited during the Cultural Revolution. The establishment of a revolutionary committee was usually accompanied by great celebrations. Congratulations poured in from other localities, and a cable of homage was sent to Mao Tse-tung. By early September 1968, the revolutionary committees had been set up in all of the 28 major administrative units. (21 provinces, excluding Taiwan, five autonomous regions, and two special municipalities, Peking and Shanghai), as well as countless political units of lower level (hsien, cities, communes, and so on), and economic, cultural, educational, and other organizations. But the establishment of the revolutionary committees merely meant the establishment of military rule over civilian life. How the transition from army to party rule was to take place is not known.

In the meantime the People's Liberation Army (PLA) was ordered by the Central Committee of the CCP to give as much support as possible to localities in promoting spring cultivation and to take over planning, procurement, and distribution in the rural areas.[5] During the summer of 1967, land and air forces of the PLA were mobilized to "assist" rural communes in harvest and cultivation. In some localities the army units supervised the peasants' work and exerted military control over areas where anti-Maoist influences had dislocated agricultural production.

Although the PLA's presence has somewhat dampened political activity in the rural areas, its ability to play a role in agriculture has been greatly impaired by lack of experience and knowledge, by contradictory orders from higher headquarters, and by conflicting ideas and loyalties within its ranks. Erroneous farming decisions resulted in serious crop damage on some communes. Farming in some areas was delayed, and as a result the most favorable time for planting was missed. Moreover, serious friction between peasants and army supervisors flared into fatal fights. On September 5, 1967, the top-level party and government authorities had to issue an order forbidding the seizure of arms, equipment, and other military supplies from the PLA.[6]

On the other hand, there were some favorable factors that tended to counterbalance the political turmoil in the countryside. Weather conditions during the Cultural Revolution and for several years afterward were, on the whole, favorable for agricultural production. Serious droughts, floods, plant diseases, and insect pests were reported in many localities; but the natural calamities in the PRC during those years seemed to be much less severe than those suffered in 1959-61. Moreover, although there was bitter criticism directed at measures adopted by Liu Shao-ch'i to stimulate the peasants' incentive after the 1959-61 crop failure—such as the extension of private plots and of free markets—these "small freedoms" apparently took root in the countryside, and have not been changed or discontinued for fear of general opposition from the peasants. In a few communes it was threatened that the peasants would be deprived of these advantages, but no action was taken.

In general it may be said that the first disorder of the Cultural Revolution gave the peasants greater freedom than they had had for several years; but after the military had consolidated its control in 1968-69, total regimentation began. For those two years or so, the policy of the early 1960s was massively criticized as the wicked capitalist line of Liu Shao-ch'i. Material incentives were denounced as the evil of all evils; and work must be done for the love of Mao Tse-tung and for the revolution. During the heyday (1969-70) of the Cultural Revolution, p'ing-chun chu-i (egalitarianism) was strictly enforced, and enormous sentiment was fostered against the idea of work for material gain. The idea of receiving or even desiring material reward for hard work was repeatedly attacked, and a massive amount of propaganda was directed toward encouraging the Maoist spirit of self-reliance and the notion that the individual should labor only for the good of society and the state, not for his own benefit. Working for material gain was denounced as "economism" and attributed to the "revisionist" influences and policies of Liu Shao-ch'i.

THE NEW ECONOMIC POLICY FOR THE COUNTRYSIDE, 1971-

In the spring of 1971, after a period of strictly military regimentation, the milder Rural Economic Policy was gradually introduced. For months radio broadcasts and newspapers provided a vast amount of evidence that the rural work-point system developed during the Cultural Revolution and exemplified by the practices at Tachai was undergoing a change. The restrictions imposed by the army and the revolutionary committees were found to be unworkable. The policy of total regimentation had led to bad management in agriculture.

The People's Daily, in an editorial on February 18, 1971, stated: "At the present stage, all people's communes must adhere to the system of 'ownership by the production team as the basic form in the three-level system of ownership of the means of production.' . . . It is also necessary to persist in the socialist principle of 'from each according to his ability and to each according to his work', . . . and handle correctly the relations between the state, the collective and the individual."[7] This and several other editorials urged the authorities to consider the response of the peasants before adopting new policies for the rural areas.

The new policy, opposing the imposition of excessive regimentation and austere military discipline upon the peasants, permitted material incentives, freedom to work private plots, and remuneration in proportion to labor performance (and not solely in proportion to political loyalty). The work-point system was modified so that higher income went to peasants who worked harder and produced more. For months Peking and provincial media spoke out against "egalitarianism" and called for the allotment of work in a way that would compensate diligence and productivity. Material incentive schemes that had been branded a "capitalist trend" during the Cultural Revolution were officially vindicated and recommended as a means of boosting peasants' initiative and production in the rural areas. Also, much was said about concern for the people's livelihood and about the rights of the peasants. Production of consumer goods, seriously neglected for years, was resumed; the commercial agencies were encouraged to serve the public better; village cadres were told to respect the peasants' rights to private plots; and the peasants' spare-time occupations and private earnings were defined more clearly. It meant, explicitly, a return to the 60-article regulations of 1961-62, which stated that land, tools, and livestock should belong to the production teams.

A document issued by the Central Committee of the CCP in December 1971 contained the following instructions: In the future permission and encouragement should be given commune members to utilize their spare time and holidays to develop family sidelines; fruit trees, bamboo, and other plants grown on self-retained plots that had been absorbed by the collectives should be returned to the individual commune members; the party's pricing policy should be correctly carried out, without overregulation of grades and prices; a rational policy of purchasing and selling should be instituted without indiscriminate alterations; the collective and the commune members should not be made to bear expenditures for the operation of the state's educational, cultural, sanitation, and other projects in the rural areas, nor for the sending of persons to the countryside; study of the labor management policies of Tachai production brigade should not be inflexibly enforced.[8]

The above measures were chiefly transitional policies aimed at restoring the national economy to its state prior to the Cultural Revolution, with an emphasis upon correcting extreme leftist tendencies in rural economic policies. By offering concessions and material incentives to the peasants, the Chinese authorities hoped to give them a new self-interest in exerting themselves for the state, and at the same time to provide a new impetus for agricultural production.

Although the new policy was a temporary concession and the fundamentals remained unchanged, the trend toward further regimentation was slowed. The absolute priority of collective agriculture was maintained, but the peasants' individual activities were encouraged.[9] The policy, however, consistently stressed that the changes put into effect must not be taken as a return to the "work points in command" situation of Liu Shao-ch'i's time. The communes were urged to draw a demarcation line between material incentives and "to each according to his work," and between justified family sideline occupations and capitalist tendencies; and to implement the new policy by strengthening education on the party's basic line and by doing away with interference from both the Left and the Right. But it was nevertheless clear that the revisions in the rural income system did, to a considerable degree, provide an incentive (in the form of a better income) for hard work. And it was also clear that political attitude now counted for less in gaining work points than it had before.

How long the new economic policy for the rural areas will continue is difficult to predict. It would seem that the combination of relaxation of centralized direction and better incentives for the peasants will have to be continued as long as the peasants resist Left pressures to further socialize agriculture by abolishing all vestiges of private property and by bringing the communes and their constituents under direct state control.

NOTES

1. Item 14, "Decision Concerning the Great Proletarian Cultural Revolution, Adopted by the Central Committee of the Chinese Communist Party on August 8, 1966," JMJP, August 9, 1966, pp. 1-2.

2. "Insist on the Correct Orientation of Going to the Rural Areas for Young Intellectuals," editorial, JMJP, July 9, 1967, p. 2.

3. "Grasp the Revolution and Promote Production," editorial, JMJP, September 7, 1966, p. 1; "More on Grasping the Revolution and Promoting Production," editorial, JMJP, November 10, 1966, p. 1.

4. JMJP, February 12, 1967, p. 1; February 20, 1967, p. 4; February 22, 1967, p. 1.

5. Ibid., February 26, 1967, p. 2.

6. NCNA, Peking, September 8 and 9, 1967.

7. "Seize the New Bumper Harvest by Following Chairman Mao's Revolutionary Line," JMJP, editorial, February 18, 1971, p. 1.

8. "Document of the CCP Central Committee, Chung-fa (1971), no. 82 [CCP Central Committee Directive on the Problem of Distribution in the Rural People's Communes, December 26, 1971], distributed with the approval of Chairman Mao." See Yeh Hsiang-chih, "An Analysis of the Current Mainland Situation," Chung-kung Yen-chiu [Studies on Chinese Communism], Taipei, Vol. VI, No. 9 (1972), pp. 1-17, English section.

9. CCP Committee of Susung hsien, Anhwei Province, "Seriously Implement the Party's Policy and Sincerely Learn from Tachai," JMJP, March 21, 1971, p. 3; Hsiang Hui, "Seriously Carry out the Party's Rural Economic Policy," Hung Ch'i [Red Flag], Peking, No. 13, 1971, pp. 30-34.

UNIVERSITY
OF BRISTOL
LIBRARY

CHAPTER

6

STATE FARMS

At the present stage there are two forms of socialist ownership of the means of production in China. One is socialist ownership by the state, that is, ownership by the whole people; the other is socialist collective ownership by the laboring masses. Socialist state ownership of the means of production includes all the mines, waterways, forests, and other natural resources, as well as some of the land. Railways, postal and telegraph offices, and banks belong to the state. The state also runs factories, farms, shops, and other enterprises and owns their products.[1]

Although the communes enforce the collective ownership system, a number of state farms have been established as "enterprises under the ownership of the whole people." The main functions of the state farms are to reclaim land, to supply the state with agricultural products for export, to serve as showcases for modern farming techniques, to supply improved seeds and animals, and to render technical assistance and guidance to the communes (APCs prior to the establishment of communes). State farms also help to solve the problems of migration, the rehabilitation and resettlement of demobilized soldiers and returned overseas Chinese, and the unemployment of urban people.

The state farms are far less significant than the communes in both the total acreage of cultivated land and agricultural production. In 1957, for example, the state farms had only 15 million mou of cultivated land, 0.89 percent of China's total cultivated land, which is estimated at 1,677 million mou. In the same year the state farms produced only 1,190 million chin of grain, 0.30 percent of China's total grain production, which was 370 billion chin. On the other hand, state farms had a much higher degree of agricultural mechanization. In 1957 they had 10,177 tractors, more than 40 percent of all the

TABLE 6.1

Statistics on State Farms and State Pastures in the PRC, Selected Years, 1949-60

	Unit	1949	1952	1957	1958	1960
Number of state farms and pastures		18	404	710	1,442	2,490
Tractors	15-hp. units	401	1,792	10,177	16,955	28,000
Harvester-combines		13	283	1,537	1,982	3,300
Trucks		28	229	3,444	4,284	n. a.
Land used for production	thousand mou	460	8,480	17,990	39,820	n. a.
Cultivated land	thousand mou	460	3,820	15,380	34,080	78,000
Reclaimed land	thousand mou	n. a.	2,240	4,060	12,430	n. a.
Number of employees	thousands	4	390	500	990	2,800
Total grain production	million chin	n. a.	n. a.	1,190	2,170	5,000

n. a. = not available.

Note: State farms and state pastures refer to those under the direct supervision of the Ministry of Agriculture prior to 1956 and the Ministry of State Farms and Land Reclamation in 1956 and thereafter, but not to those administered by provinces, autonomous regions, municipalities, and other local authorities.

Sources: All figures for 1960 and the figure for grain production in 1957 are from Wang Chen, "Strengthening the Construction of State Farms," Hung Ch'i [Red Flag], Peking, No. 7, 1961, pp. 1-7. The figure for grain production in 1958 if from "March Forward Bravely on the Land Reclamation Front," editorial, Chung-kuo Nung-k'en [China Reclamation], Peking, No. 18, 1959, pp. 1-2. All other figures are from State Statistical Bureau, Ten Great Years (English ed.; Peking, 1960), p. 134

24,629 tractors operated in the PRC in that year (see Tables 6.1 and 6.2). Another source indicated that in 1964 the total acreage of cultivated land on the state farms was 290 percent of that in 1957 but still was "only a fraction of the total acreage of cultivated land for the country as a whole." In the same year state farms owned 32 percent of the country's tractors; 50 percent of the mechanical farm tools; 82.5 percent of the combines; 68 percent of the heavy-duty trucks; "several hundred thousand units" of irrigation and drainage equipment, power-generating equipment, and processing equipment; and several hundred repair shops of various sizes.[2]

Large state farms generally are under the direct supervision of the Ministry of State Farms and Land Reclamation, established in May 1956; smaller farms are administered by provincial, municipal, special district, and hsien authorities. Most of the large state farms are in one of the four major reclamation areas established by the ministry. The Sinkiang reclamation area, the largest of the four, is operated by the Sinkiang Production-Construction Corps, under the Sinkiang Military Area Command of the PLA, with a total sown acreage in 1961 of 8 million mou of corn, spring wheat, cotton, soybeans, rapeseeds, and sugar beets. The other three areas, operated by the Northeast General Bureau of Land Reclamation, are the Mutankiang and Hokiang reclamation areas in Heilungkiang Province and the Hulunbar reclamation area in Inner Mongolia. Each of these areas sowed some 4 million mou of grains and soybeans in 1961.[3]

Although data on state farms in recent years are sketchy, their general situation and problems appear to be essentially the same as they were in the 1950s and the 1960s. At a provincewide symposium of state-owned agricultural and animal husbandry enterprises, convened in Chouchih, Shensi Province, in May 1975, the participants were told that although the proportion of the enterprises in the entire rural economy was "still not great," they occupied an important position in China's socialist agriculture. The state-owned enterprises were urged to perform their functions in accordance with the Tachai spirit.[4]

The People's Daily, in a short commentary, reported that state-run farms in Tibet had made great achievements through diligence and frugality, although some people had said that these farms were difficult to run and that losses had been tremendous.[5] State farms in the Ninghsia-Hui Autonomous Region and 45 of them in Hupeh Province also had run at a loss until recently, according to the People's Daily. The paper urged the state farms to run more efficiently by learning from Tachai. "There is no truth in the assertion made by some people on the state farms that loss of money is inevitable," the paper said.[6]

Table 6.3 summarizes the development of socialist agriculture in the PRC in 1950-65 and 1972, in terms of the number of farm units and the percentage of farm households participating in various stages

TABLE 6.2

Number and Percentage of Tractors Operated by State Farms in the PRC, 1949-65

Year	Number of Tractors Operated by All Farms	Number of Tractors Operated by State Farms	Percent of Tractors Operated by State Farms
1949	401	401	100.0
1950	1,286	n.a.	n.a.
1951	1,410	n.a.	n.a.
1952	2,006	1,792	89.3
1953	2,719	1,801	66.2
1954	5,061	2,766	54.7
1955	8,094	4,036	49.8
1956	19,367	7,242	37.4
1957	24,629	10,177	41.3
1958	45,330	16,955	37.4
1959	59,000	n.a.	n.a.
1960	79,000	28,000	35.5
1961	99,000	n.a.	n.a.
1962	100,000	n.a.	n.a.
1963	115,017	n.a.	n.a.
1964	123,000	39,360	32.0
1965	154,161	n.a.	n.a.

n.a. = not available.

Sources: Number of tractors operated by all farms—1949-58, State Statistical Bureau, Ten Great Years (English ed.; Peking, 1960), p. 135; 1959, Ch'en Cheng-jen, "Speed up the Technical Transformation of Agriculture," Hung Ch'i [Red Flag], Peking, No. 4, 1960, pp. 4-10; 1960 Chang Feng-shih, "Produce Suitable Agricultural Machinery to Support Agriculture," Kung-jen Jih-pao [Daily Worker], January 18, 1962, p. 1; 1961, JMJP, February 5, 1962, p. 1; 1962, "Actively and Systematically Bring About the Technical Transformation of Our Country's Agriculture," editorial JMJP, November 9, 1962, pp. 1-2; 1963, News release, China News Service, Canton, January 5, 1964; 1964, based on the statement "Compared with 1957, the number of tractors employed in agriculture increased by four times in 1964," in Chou En-lai's "Report on the Work of the Government," delivered at the First Session of the Third National People's Congress, December 21-22, 1964 (NCNA-English, Peking, December 30, 1964), 1965, based on the statement "Agricultural machine stations and state farms together owned 7.96 times as many standard units of tractors as they did in 1956," in "Paving a Way to Agricultural Mechanization in China—A Visit to the Agricultural Mechanization Hall of the National Agricultural Exhibition," JMJP, April 13, 1965, p. 1.

Number of tractors operated by state farms—1949, 1952, 1957, and 1958, State Statistical Bureau, Ten Great Years (Peking, 1960), p. 134; 1953-56, "Great Achievements of State Farms in the Past Few Years," Chung-kuo Nung-k'en [China Reclamation], Peking, No. 4, 1957, pp. 24-25; 1960, Wang Chen, "Strengthen the Construction of State Farms," Hung Ch'i, No. 7, 1961, pp. 1-7; 1964, based on other two columns of the table.

Percentage of tractors operated by state farms—1964, Shang Chih-lung and Ma Ching-p'o, "Fifteen Years of Agricultural Mechanization of State Farms," Nung-yeh Chi-hsieh Chi-shu [Technology of Agricultural Machinery], Peking, No. 11, 1964; other years, based on other two columns of the table.

TABLE 6.3

Development of Socialist Agriculture in the PRC, 1950-65, 1972

Year	Mutual-Aid Teams	Primary APCs	Advanced APCs	Rural Communes	State Farms	Percent of Farm Households in Socialist Agriculture
1950	2,724,000	18	1	—	n.a.	10.7
1951	4,675,000	129	1	—	n.a.	19.2
1952	8,026,000	3,634	10	—	404	40.0
1953	7,450,000	15,053	15	—	n.a.	39.5
1954	9,931,000	114,165	201	—	n.a.	60.3
1955	7,147,000	633,213	529	—	n.a.	64.9
1956	—	682,000	312,000	—	n.a.	96.3
1957	—	72,022	680,081	—	710	97.0
1958	—	—	—	26,578	1,442	99.1
1959	—	—	—	24,000	n.a.	n.a.
1960	—	—	—	n.a.	2,490	n.a.
1961	—	—	—	26,000	2,500	n.a.
1962	—	—	—	n.a.	n.a.	n.a.
1963	—	—	—	n.a.	n.a.	n.a.
1964	—	—	—	74,000	n.a.	n.a.
1965	—	—	—	n.a.	2,000	n.a.
1972	—	—	—	50,000	n.a.	n.a.

n.a. = not available.

— = not applicable.

Source: Adapted from Frederick W. Crook and Linda A Bernstein, Agriculture in the United States and the People's Republic of China, 1967-71, Foreign Agricultural Economic Report No. 94 (Washington, D.C.: Economic Research Service, U.S. Department of Agriculture, 1974), p. 8, Table 1.

of the socialist transformation of agriculture—mutual-aid teams, APCs, and communes. It shows the rapid pace of cooperativization of Chinese agriculture in 1953-56. By the end of 1956, nearly all (96 percent) farm households were in the APCs. The table also shows that the number of communes almost doubled, from 26,578 in 1958 to an estimated 50,000 in 1972, as a result of the splitting up of many communes that were found to be too large for efficient management.

Although the number of state farms increased rather rapidly, from 404 in 1952 to 2,000 in 1965, their cultivated land and grain output remained insignificant compared with those of the communes. The Chinese Communists believe that the system of collective ownership will eventually be replaced with ownership by the entire people, but they also admit that this will be a "fairly long process of gradual development."[7] Since the early 1950s there have been no signs from the authorities that they intend to increase the rate of transition from communes to state farms. Although ownership by the whole people has barely started, the Chinese leadership generally considers that the socialist transformation of agriculture has been completed.

NOTES

1. The Constitution of the People's Republic of China, adopted by the Fourth National People's Congress at its first session, January 17, 1975, Article 6; Workers' Political Economy Study Group of the Shanghai No. 4 Hosiery Factory and Cheng Ching (of Futan University's political economy department), "Public Ownership of the Means of Production in China," Peking Review, No. 51, 1972, pp. 5-7.

2. Shang Chih-lung and Ma Ching-p'o, "Fifteen Years of Agricultural Mechanization on State Farms," Nung-yeh Chi-hsieh Chi-shu [Technology of Agricultural Machinery], Peking, No. 11, 1964.

3. NCNA, Peking, July 9, 1961.

4. Shensi Provincial Service (in Chinese), Sian, May 21, 1975; translated in FBIS Daily Report, PRC, May 22, 1975, p. M-1.

5. JMJP, May 24, 1975, p. 2.

6. Ibid., October 20, 1975, p. 3.

7. Workers' Political Economy Study Group of the Shanghai No. 4 Hosiery Factory and Cheng Ching, op. cit.

PART II

TECHNICAL TRANSFORMATION OF AGRICULTURE

INTRODUCTION

With regard to land utilization, a country's agricultural production depends upon the area under cultivation and the yields per unit of the existing cultivated land. Throughout much of the world's history, man has increased the supply of food and other agricultural products mainly by expanding the areas under cultivation, and yields have remained rather constant at low levels. Only over the past two centuries have average world yields begun to make consistent gains.

In the years since 1950 especially, the four geographic regions classified as developed—North America, Western Europe, Oceania, and the Soviet bloc—have been largely or entirely dependent on yield increases for additional output of grain and other principal crops. Of the three less developed geographic regions, much of South America and Africa have continued to be more dependent on additions to the cultivated area than on higher yields. But during the 1960s the densely populated Asian countries, with their diminishing land expansion potentials, have become more and more dependent on raising yields for increasing agricultural production. It is generally believed that even in some countries of Africa and Latin America the greater relative dependence on addition of new land is not likely to continue much longer, and raising yields may be the better means of increasing their agricultural output.[1]

Like other densely populated, less developed countries, China has nearly exhausted the supply of new land that can readily be brought under cultivation. It therefore must look to raising yields for most of the additions to the supply of food and other agricultural products. In fact, for centuries intensive farming has been practiced in most agricultural regions of the country. Extensive use of water, control of soil erosion, fertility maintenance, crop rotation, and control of weeds have largely accounted for the high yields of many crops. In more recent decades other measures, including plant breeding, control of insects and diseases, and more extensive use of chemical fertilizers, have been adopted to further raise yields. As a result of these efforts, agricultural production in general during the 1920s and 1930s improved rather steadily, although not sufficiently to meet the increasing needs. In 1936 China had its largest agricultural production in many years, and this was attributed to the general agricultural improvement measures as well as to the effect of political stability and favorable weather.

However, the improvement was abruptly interrupted by the Sino-Japanese War (1937-45) and the civil strife that followed. Military operations affected large sections of the population and hindered normal cultivation of the land in many areas. More than 70 million mou (11.7 million acres) of farmland were ravaged by the war, and some 17.5 million farmers suffered directly or indirectly. Another 14 million mou (2.3 million acres) of farmland were devastated by the flood of the Yellow River in Honan Province, affecting more than 3 million persons. (The flood developed when the dike on the right bank of the river near Chengchou was broken intentionally in 1938 by the Nationalist government in order to slow the advance of the Japanese armies.) Irrigation systems for more than 10 million mou (1.7 million acres) of farmland were destroyed or laid waste. Transportation and communication facilities in many sections of the country were damaged. Not only did agricultural production again decline to a low level, but industries closely related to agriculture—such as cotton spinning, flour milling, and silk reeling—suffered heavily. Agricultural trade was sharply reduced, and a number of agricultural research and experiment projects were discontinued.

During the Sino-Japanese War most of the Chinese government's activities were wartime emergency measures. Following the cessation of the hostilities, important steps were taken to reconstruct agriculture with the assistance of the United Nations Relief and Rehabilitation Administration and a survey of problems relating to the development of China's agriculture was made in 1946-47 by a mission composed of American and Chinese specialists appointed by the two governments. Some progress was made but, because of the immensity of the problems as well as the depleted resources and insufficient personnel, much remained to be done. Moreover, the conflict between the Nationalist government and the Communists deepened, and agricultural rehabilitation was badly affected by the civil strife.[2]

Thus, when the Communist regime took over Mainland China in 1949, it did not inherit a bountiful agriculture. There were numerous problems to be solved. Some of them were urgent and others of a long-range nature. The task required large amounts of materials and services. The new government's first objective in economic planning was industrialization, however, and agriculture was not given as much attention as it should have received. During the Economic Recovery Period (1949-52) programs and measures adopted for the improvement of agricultural production were mostly piecemeal and of a regional or local nature. Agriculture was included as a major sector of the national economy in the First Five-Year Plan (1953-57) and the Second Five-Year Plan (1958-62), but its significance was much overshadowed by the emphasis on industrial development.

It was not until 1960, the second of three consecutive years of bad harvest, that the policy of "developing agriculture as the foundation of the national economy and industry as the leading factor" was adopted. Since then the order of priority among the major sectors of the national economy has been agriculture, light industry, and heavy industry. One of the principal measures to implement the new emphasis on agriculture was for industry to provide as much aid as possible to agriculture. In fact, aid to agriculture was to be the most important way in which industry would fulfill its leading role in the national economy. Industrial development was to be primarily the development of machinery, equipment, and materials that would improve agricultural production.[3] More specifically, the most important objective of industrial aid to agriculture was to speed up the technical transformation of agriculture, for which goals and measures were provided in the National Program for Agricultural Development (NPAD), 1956-67. The purposes, contents, and implementation of the program will be discussed in the following chapters.

NOTES

1. Food and Agriculture Organization of the United Nations, The State of Food and Agriculture, 1968 (Rome, 1968), p. 76.
2. Government of the Republic of China, "Annual Progress and Program Report on Food and Agriculture for China, 1947," submitted to the Conference of the Food and Agriculture Organization of the United Nations (Washington, D.C., August 1947). (Mimeographed.)
3. Leslie T. C. Kuo, "Industrial Aid to Agriculture in Communist China," International Development Review, Washington, D.C., Vol. IX, No. 2 (1967), pp. 6-10, 29.

CHAPTER

7

NATIONAL PROGRAM FOR AGRICULTURAL DEVELOPMENT

The National Program for Agricultural Development (NPAD), 1956-67, was described in its preamble as "a program of endeavor to develop agricultural production in our country and to raise the living standard of the peasants and the people as a whole during the period from the First Five-Year Plan to the Third." (See Appendix for a complete English translation of the program.) The core of the program consisted of the goals and measures to expand the area under cultivation and to raise yields per unit of existing cultivated land. In the 12 years beginning in 1956, the area cultivated by state farms was to be increased from the 1955 figure of more than 13 million mou to about 100 million mou (Article 16). The average annual yields of grain crops in the three principal grain-growing regions were to be raised from 150 to 400 chin per mou, from 208 to 500 chin per mou, and from 400 to 800 chin per mou, respectively; and the average annual yield of ginned cotton was to be raised from the 1955 figure of 35 chin per mou (the national average) to 40, 60, 80, or 100 chin per mou, depending on local conditions (Article 2).

The most important measures for attaining these goals, as prescribed in the program, included the following:

1. Construction of water conservation projects
2. Expansion of the fertilizer supply
3. Improvement of old-style farm tools and popularization of modern farm tools
4. Extension of the use of the best and most suitable plant varieties
5. Expansion of multiple-cropping areas
6. Planting of more high-yielding crops

7. Adoption of the best farming methods
8. Improvement of the soil
9. Conservation of soil and water
10. Protection and breeding of draft animals
11. Extermination of insects, pests, and plant diseases
12. Reclamation of wasteland.

For some of these measures, the program also set goals to be fulfilled by the end of 1967, including the expansion of paddy fields and irrigated land from 390 million mou in 1955 to 900 million mou in 1967; the raising of the capacity of the irrigation facilities to control droughts to between 30 and 50 days and, in places suitable for double-crop paddy fields, to between 50 to 70 days; the annual production of 5 to 7 million tons of chemcial fertilizers by 1962, and 15 million tons by 1967; and the elimination of the most harmful insect pests and plant diseases within the 12 years 1956-67.

Although the 40-article program was primarily a blueprint for agricultural development in the PRC, it also covered such problems as general welfare, public health, education, transportation, communications, commerce, finance, household management, the eradication of rats, flies, mosquitoes, and bedbugs, and the encouragement of birth control. The preamble pointed out that these and many other tasks required the participation of both rural and urban people, and that they could be effectively carried out only with the coordinated efforts of the towns and the countryside.

The contents and goals of the program have been a matter of dispute among party leaders. The program was first proposed by Mao Tse-tung in 1955 during the "high tide of rural socialism." It was submitted in draft form by the Political Bureau of the Central Committee of the CCP to the Supreme State Conference, which adopted it on January 25, 1956. But, in the second half of 1956 and the beginning of 1957, there was strong opposition to the program, which many Communist leaders considered to be overzealous and unrealistic. Thus, the Eighth National Congress of the CCP, held in September 1956, did not approve the program. However, following the Congress, Mao Tse-tung relied on the support of the provincial secretaries of the party to overcome his opponents and make the goals of the program even more ambitious. As a result, the draft was first amended at the Third Plenary Session of the Eighth Central Committee of the CCP on October 9, 1957, and again at the Second Session of the Eighth National Congress of the CCP on May 23, 1958.[1]

Despite the revisions Mao Tse-tung and his followers did not consider the goals prescribed in the program to be high enough. During the Leap Forward campaign the program was actually sidelined. In the "Decision on the Question Concerning the Establishment of Rural

Communes," adopted by the Central Committee of the CCP on August 29, 1958, the nation was urged to increase agricultural production by "two, several, ten and a dozen times."[2] When the Draft Program of the National Economy for 1959 was adopted by the Seventh Plenary Session of the Eighth Central Committee of the CCP on April 2-5, 1959, the grain production target for that year was set at 1,050,000 million chin (525 million metric tons), which was one-third more than the target for 1967 (720,000 million chin, or 360 million metric tons), prescribed in the NPAD.[3]

But the economic dislocation and agricultural crisis that were caused by the Leap Forward campaign and aggravated by the natural calamities in 1959 and 1960 made it necessary to bring the NPAD up for discussion again, after nearly two years of oblivion. Most of the production targets proposed by Mao Tse-tung and others had proved unattainable; and at the Second Session of the Second National People's Congress on April 10, 1960, the more conservative first revised draft of the program was adopted with one insignificant change: Article 27 was amended so that bedbugs took the place of sparrows among the four pests to be exterminated, because it had been discovered that sparrows, unlovable as they are, are indispensable as the enemies of certain insects.[4] The first revised draft thus became the final version of the program, superseding the second revised draft, which was never published.

NOTES

1. JMJP, October 10, 1957, p. 1; T'an Chen-lin, "Strive for the Fulfillment, Ahead of Schedule, of the National Program for Agricultural Development," JMJP, April 9, 1960, p. 1.
2. Hsin-hua Pan-yueh-k'an [New China Semimonthly], Peking, No. 18, 1958, pp. 1-2.
3. Ibid., No. 8, 1959, p. 1.
4. JMJP, April 12, 1960, p. 1.

CHAPTER

8

IMPLEMENTATION OF THE PROGRAM

Agricultural research, education, and extension are three of the essential services that a government must provide for a country's agricultural development. Research evolves new ideas and new techniques, education provides trained personnel, and extension assists farmers in putting into practice the products of research. The three services are complementary and mutually dependent. The effectiveness of the government's effort to improve agricultural production depends on the organization and administration of these services. A brief account of governmental organization and functions for agricultural development, as well as overall policies and guiding principles of agricultural research, education, and extension, in the PRC is given here as a background for discussing major aspects of the technical transformation of agriculture.

Like other economic activities, agricultural development in the PRC is completely controlled by the government and, in turn, by the Communist party. Under the State Council, the highest executive body of the PRC government, are the Office for Agriculture and Forestry and several ministries that perform agricultural functions. Prior to the Cultural Revolution the ministries included Agriculture, Forestry, Food, Marine Products, State Farms and Land Reclamation, Water Conservation and Electric Power, Agricultural Machinery (renamed Eighth Ministry of Machine-Building Industry in 1965), and the Second Ministry of Machine-Building Industry, which was in charge of semi-mechanized agricultural implements.

In keeping with the process of streamlining the central government structure after the purges of the Cultural Revolution, the ministries of Agriculture, Forestry, Marine Products, and State Farms and Land Reclamation were merged in 1970 to form the Ministry of Agriculture and Forestry. The First Ministry of Machine-Building

Industry was expanded in 1970 by taking over the functions of the former Eighth Ministry of Machine-Building Industry, which had been abolished. Prior to the Cultural Revolution there were several commissions, bureaus, and special agencies under the State Council, each of which performed functions related to several fields, including agriculture. These included the State Capital Construction Commission, the State Statistical Bureau, the State Scientific and Technological Commission, the Agricultural Bank of China, and the Central Meteorological Bureau.

Above all these agencies is the Rural Work Department, which is directly under the CCP and makes paramount decisions on agriculture. For the supervision of work in communes, the Political Department of Agriculture and Forestry was created in August 1965, under the Central Committee of the party.[1] Inevitably there is much duplication and overlapping in the functions and operations of these organs under the weighty makeup of the central government. The structures of the provincial, hsien, commune, and other local government units are equally complicated.

Since the Cultural Revolution all these central, provincial, and local government units have been run by the respective revolutionary committees; and in each committee the party nucleus is the core of the leadership.[2] Obviously all agricultural activities are under tight party control.

AGRICULTURAL RESEARCH

Before the Communist take-over agricultural research in China at the national level centered in (1) the National Bureau of Agricultural Research at Nanking, dealing mainly with crops; (2) the National Bureau of Animal Husbandry Research at Nanking; (3) the National Bureau of Forestry Research at Nanking; and (4) the National Bureau of Fisheries Research at Shanghai. A National Bureau of Agricultural Economics Research was planned shortly after the Sino-Japanese War. All of these bureaus were directly under the Ministry of Agriculture and Forestry; and each had experiment stations, field stations, and laboratories. In addition, a number of other agencies in the Ministry of Agriculture and Forestry conducted agricultural research to some degree. Research also was done by the national colleges of agriculture.

At the provincial level there was a great array of agricultural research institutes, agricultural experiment stations, seed- and animal-multiplication stations, demonstration farms, and other institutions for agricultural improvement. In general these institutions were understaffed, poorly equipped, and inadequately financed. Most of them conducted simple tests of fertilizers, pesticides, and plant varieties. A few occasionally attempted more basic research, but little of their work was systematically organized.[3]

During the early years of the Communist regime, agricultural research in the PRC was conducted for the most part by several agencies under the Ministry of Agriculture and by a number of agricultural colleges. The Chinese Academy of Sciences, which was entrusted with the task of promoting and coordinating scientific research work generally, also conducted some research in agronomy, entomology, plant pathology, forestry, and water conservation.

The Chinese Academy of Agricultural Science was founded in Peking on March 1, 1957, by the Ministry of Agriculture. Ting Ying, a rice specialist, was elected its first president.[4] As the focal point for the development of agricultural research, this academy, in addition to carrying out its own program, coordinates the research work of nearly 30 agricultural colleges and more than 200 other organizations engaged in agricultural experimentation and research. The Chinese Academy of Agricultural Science also cooperates with related institutes of the Chinese Academy of Sciences.

The Chinese Academy of Agricultural Science has established more than 20 research institutes, each specializing in a major field of agriculture, and a number of regional and provincial research institutes. Many branch institutes and experiment stations also have been set up in hsien and communes throughout the country. Despite the fact that policy making and the control of agricultural research operations are highly centralized in the government, the research institutions are widely distributed throughout the country. Although many of the research institutes under the Chinese Academy of Agricultural Science are located in Peking, the national capital, other institutes have been set up in areas where there is substantial production of the crop or animal in question. Among the institutes that have been established in Peking are those for plant breeding and cultivation, soil and fertilizers, agricultural uses of atomic energy, plant protection, vegetables, animal husbandry, apiculture, agricultural mechanization, agricultural meteorology, agricultural economics, and horticulture. Institutes located in other cities include those for sericulture (Chinkiang, Kiangsu Province), agricultural heritage (Nanking), tea (Hangchou, Chekiang), pomology (Hsincheng, Liaoning), cotton (Anyang, Honan), hemp (Yuanchiang, Hunan), potatoes (Shuchien, Kiangsu), marsh gas (Fuyang, Anhwei), Chinese veterinary medicine (Lanchou, Kansu), citrus (Chungking), and tobacco (Tsinan, Shantung). Institutes for the following fields also have been mentioned, although their precise locations are not known: veterinary science (Western), horses, oxen, water buffaloes, hogs, sheep, hide and fur animals, poultry, oil-bearing crops, oak silkworm culture, and farm irrigation.

In addition, there are six regional research institutes under the Chinese Academy of Agricultural Science: for northeast China, northwest China, east China, south China, north China, and central China. There are also provincial research institutes in Chekiang, Hupeh, Shensi, Shantung, Shansi, Anhwei, Kwangtung, Fukien, and Kiangsu. Many hsien, municipalities, and communes also have established institutes and other research and experimental facilities to further extend the organizational network for agricultural research.

Among the host of slogans spawned by Mao Tse-tung, "walking on two legs" has the peculiar merit of apparently fitting any situation that confronts the PRC. Many interpretations have been poured into this mold. At the beginning of the Leap Forward campaign in 1958, for example, the slogan stood for the simultaneous development of backyard primitive furnaces and of modern metallurgy with Soviet assistance. When the primitive furnaces petered out in early 1959, the slogan was made to cover the parallel development of heavy and medium metallurgy.

In the technical transformation of agriculture, "walking on two legs" means numerous things. Modern mechanization of farming is carried out simultaneously with the improvement and innovation of native tools and implements. The technique of acupuncture and other methods of native veterinarians are used together with the findings of modern veterinary science. In improving the soil the use of both chemical and organic fertilizers is promoted. In plant protection chemical pesticides as well as "native agricultural drugs" are used. In water conservation large reservoirs and dams are constructed side by side with small canals, ditches, and wells.

Scientists have gone to the villages to learn about the farmers' ways, problems, and common beliefs. At the same time farmers are encouraged to participate in experimentation, the development of new plant varieties, the designing of improved tools and implements, the survey of soils, and other activities. Thus the scientists and the masses are brought close together. Communist leaders recognize that it will take a long time to develop and apply modern agricultural science on a broad basis in China; and they believe that agricultural production can be increased at a higher speed by using the combined methods of old and new, native and foreign, scientific and primitive. At the same time the emphasis on having research workers go to the countryside to work with the peasants appears to be aimed at providing a means for indoctrinating the scientists with Communist principles and practices and thus making it more certain that they remain an integral part of the Communist pattern.

In addition to conducting laboratory and field experiments, the Chinese Academy of Agricultural Science has another major function that it carries out simultaneously: "collecting the past experiences of the masses." During the latter half of 1958, a majority of the academy's technical staff members lived and worked with farmers in the villages from time to time, in order to discuss the farmers' production experiences with them and at the same time to give the farmers any assistance they might need. In 1959 more than 360 centers were established for this purpose. On the basis of information collected from the farmers, publications dealing with rice, cotton, wheat, fruits, mulberries, animal diseases, plant protection, fertilizers, soil conservation, and other subjects have been prepared.[5]

A much-publicized example of "walking on two legs" was Ch'en Yung-k'ang, a farmer of Kiangsu Province, who in 1950 was selected as a national model farmer and in 1958 was appointed to the research staff of the East China Agricultural Research Institute of the Chinese Academy of Agricultural Science. Ch'en has to his credit a high-yielding rice strain that he bred from a rice plant selected 20 years earlier in his rice fields. Based on his observation of the growth, shape, and subtle color changes of the leaves of the rice plants on a few mou of land, Ch'en maintained that the leaves of late rice change from dark green to lighter green, and vice versa, six times before the harvest. In his opinion manure and water should be so applied as to make the change of color occur at the proper time, in order to lead to sturdy growth and high yield. He called this the theory of san huang san hei (three yellow and three black).

Ch'en's skill in rice growing is said to have further improved while he was collaborating with scientists of the Chinese Academy of Agricultural Science. Working together, they achieved yields of about 1,000 chin of rice per mou on their experimental plots, and the highest yield reached 1,240 chin of upland rice per mou.[6]

Some supported Ch'en's theory; others maintained that the change in leaf color was a common phenomenon that had nothing to do with the growth of rice and that Ch'en's harvests were not sufficient basis for drawing scientific conclusions.[7] In any case committees for popularizing Ch'en Yung-k'ang's rice-growing methods were set up in 1963 by all governments and communes in the Soochow District, Kiangsu Province. His methods were applied on experimental farms totaling more than 6,000 hectares in thousands of villages, all of which had better harvests in 1963 than in 1962, according to the People's Daily.[8]

Another example of "walking on two legs" is Pei Yao-ching, a peasant who learned Chinese native veterinary methods from his grandfather and his uncle. Pei has compiled or rewritten several important works on traditional Chinese veterinary practices. In 1950 he was appointed a veterinarian and later deputy chief of a veterinary hospital in central Shansi. In 1963 he became a special research worker at the Research Institute of Traditional Chinese Veterinary Science under the Chinese Academy of Agricultural Science.[9]

As a move to propagate the old agricultural techniques, the Research Institute of Agricultural Heritage was established at Nanking in 1958 by the Chinese Academy of Agricultural Science. Under the auspices of the institute, more than 30 ancient agricultural publications have been reprinted. Most of these publications are from a collection of old works on agriculture that formerly was maintained by the University of Nanking. These include Shou-shih T'ung-k'ao [A Comprehensive Reference Book on Agriculture], an imperial compilation by Chang T'ing-yu and others, originally published in 1742 in 78 volumes containing quotations on agriculture from old Chinese literature; Ch'i-min Yao-shu [Important Techniques of the Common People], a book on the methods of growing and processing agricultural products by Chia Szu-hsieh, originally published in the sixth century; Fan-sheng-chih Shu [A Book by Fan Sheng], a work on agricultural techniques, published in the first century B.C.; Nung Shu [Book on Agriculture], by Wang Chen (fourteenth century); and works on Chinese veterinary therapeutics and prescriptions, Chinese methods of cultivating various plants, and other agricultural topics. Some of these reprints have been issued with interpretations and suggestions as to how some of the old farming practices may be adapted for Chinese agriculture today.[10]

The Chinese Communists have followed the principle of subservience of scientific research to the requirements of the state. The pursuit of scientific research generally is based on the country's political, military, and economic development plans rather than on the interests of individual scientists. In agriculture the policy has aimed at agricultural research for the sake of agricultural production. Research and experimentation with the objective of immediate increase in agricultural production have been emphasized, although more basic or theoretical studies have not been entirely neglected.

Since the principle of subservience requires centralized direction and control to ensure fulfillment of plans, virtually all research institutions and projects are under the supervision of party cadres, most of whom are not scientists. Science that presupposes slow, painstaking work and fails to show immediate results in helping economic development is not appreciated by the cadres. On the other hand, many of the scientists strongly feel that they should devote themselves to research on basic and theoretical problems rather than give their time to problems of economic development and elementary scientific training.

Latent differences between scientists and the Communist party broke into a bitter struggle during the movement of "Letting a Hundred Flowers Bloom and a Hundred Schools Contend" in 1956. As the outcome of the controversy, the PRC's political planners and policy makers took firmer control in the direction and administration of scientific research. During the Leap Forward campaign, many scientists were seized by the mania for speed under the slogan of "producing more, better, faster, and economically." As pointed out by a correspondent of the People's Daily with reference to a study on the deterioration of potatoes conducted by the Research Institute of Microbiology, the scientists "had to abandon the original basis of study and tried to work out a simple method which would put a stop to the deterioration at one blow. For the sake of speed, they sacrificed the verification of results. This demand for speed was in obvious contradiction to scientific work. The only result of it was waste of manpower and waste of time with an unfortunate influence on research work."[11]

AGRICULTURAL EDUCATION

Prior to the Communist take-over, China had 25 national colleges of agriculture (21 Chinese and 4 missionary) and nearly 20 provincial and private agricultural colleges. These higher educational institutions provided training in agriculture at the undergraduate level, and by the end of 1946 they had graduated more than 6,000 students. Some of the graduates received further training abroad. In addition to these colleges, there were also a number of agricultural vocational schools of intermediate level. Although some of these institutions, such as the University of Nanking, did outstanding work, the system of agricultural education as a whole was inadequate. Moreover, many of the graduates were not in positions where they could best utilize their training.[12]

Large numbers of technical workers of all levels were reported to have been trained in agriculture since 1949. As a result of the drastic reorganization of China's educational system, which began in the latter part of 1950, virtually all of the original institutions of learning have been merged or abolished and many new ones have been established. The number of higher educational institutions in agriculture increased from 23 (1 university and 22 colleges) in 1957 to 42 (7 universities, 26 colleges of agriculture, and 9 colleges of agricultural mechanization) in 1966. Facilities as well as faculty varied widely among these universities and colleges. For example, in addition to trained personnel, many "native experts" (peasants who are experienced in local farming) have been appointed to give instruction and lectures at some colleges.[13]

Table 8.1 shows the annual number of graduates from institutions of higher learning in agriculture, as compared with the number of graduates in all fields, in 1949-58, 1961, and 1962. The annual number of agricultural graduates represented only 4.8 to 8.2 percent of the total number of graduates in all fields during the period 1949-61, partly because under the industrialization program there was a greater demand for technical personnel in nonagricultural fields. The

TABLE 8.1

Number of College Graduates in the PRC, 1949-58, 1961, 1962

Year	Total Graduates	Agricultural Graduates: Number	Agricultural Graduates: Percent of Total Graduates
Peak year before 1949	25,000	2,064	8.3
1949	21,000	1,718	8.1
1950	18,000	1,477	8.2
1951	19,000	1,538	8.1
1952	32,000	2,361	7.4
1953	48,000	2,633	5.5
1954	47,000	3,532	7.5
1955	55,000	2,614	4.8
1956	63,000	3,541	5.6
1957	56,000	3,104	5.3
1958	72,000	3,513	4.9
1961	162,000	10,000	6.2
1962	170,000	20,000	11.8

Sources: Figures for the peak year before 1949 and for 1949-58 are from State Statistical Bureau, Ten Great Years (English ed.; Peking, 1960), pp. 194, 196. Figures for 1961 and 1962 are from JMJP, August 4, 1961, p. 4, and August 29, 1962, p. 1, respectively.

proportion of agricultural graduates rose to 11.8 percent in 1962, following the adoption of the policy of "developing agriculture as the foundation of the national economy."

The training of postgraduates has progressed very slowly in all fields. Before the Sino-Soviet split, such training was carried out mainly by sending students to the Soviet Union and other Communist countries. In recent years Communist China has increasingly relied on its own efforts in training postgraduates, but has admitted that the program is far from adequate, both quantitatively and qualitatively.[14]

The Chinese Academy of Sciences was the first institution to recruit postgraduate students. In 1955 the academy offered a four-year curriculum to 50 candidates for the degree of Associate Doctor of Science.[15] The following year four institutes of the Chinese Academy of Sciences recruited 361 postgraduates for advanced studies in 18 specialized courses.[16] Since then the academy has continued such recruitment every year.

The recruitment of postgraduates by the institutions of higher education began in 1956, when for the first time several universities and colleges offered a four-year curriculum for advanced degrees. In the first two years the candidates were required to study dialectical materialism and historical materialism, foreign languages, and basic specialized subjects. The last two years were spent mainly in the preparation of theses.[17]

In recent years both the number of institutions recruiting postgraduates and the number of postgraduates enrolled have increased considerably. In 1964 more than 180 institutions of higher education and research organs recruited 1,280 postgraduates, as compared with 160 institutions and 800 postgraduates in 1963.[18] No data are available to indicate how many of these were in agriculture.

Below the college level, in addition to the regular middle schools there are a number of technical middle schools that give short courses on various subjects, including agriculture. The number of students enrolled in these technical middle schools increased from 230,000 in 1949 to 1.47 million in 1958, and thc number of graduates lncreased from 72,000 to 191,000 during the same period.[19]

Since early in 1958 more than 200,000 agricultural middle schools have been established in the communes as the most important institutions for the training of intermediate technical personnel in agriculture. Boys and girls 13 to 16 years of age who work in the communes are selected to spend half of their time in the schools, that is, six months each year for three years. The level is equivalent to that of junior middle schools, and the curriculum includes basic agricultural techniques as well as politics, Chinese, and mathematics. Emphasis is given to the special farming problems and crops of the particular locality. Most of these schools arrange their schedules so that the students can work full time in their communes during the busy farming season. At the end of 1964, some 400,000 students had graduated from the agricultural middle schools in Kiangsu, the first and model province for such institutions, and another 200,000 were still studying in the schools. The Communist government hopes that these students will form a major force of technical workers on the farms when China's agriculture becomes more modernized.[20] But press reports indicate that the academic standards of these agricultural middle schools, as well as their accommodations and facilities, are generally low. Moreover, many cadres in charge of the schools, teachers, and students are not enthusiastic about their work.[21]

Since 1964 the system of combining study with farming has been expanded to educational institutions of all levels in the rural areas, in pursuance of the Chinese Communist party's policy that education must serve proletarian politics and be linked with productive labor. By September 1965 the system had been adopted by 37 of the 66 agricultural colleges or institutes in the country with 11,800 students, or 15 percent of the total enrollment, and by 220 of the 307 agricultural middle schools with 41,600 students, or 52 percent of the total enrollment.[22] At the same time the whole country had a total of 17 million children attending the part-farming and part-study primary schools.[23] In 1965, 90 percent of middle schools and primary schools and 80 percent of their total enrollments were in the countryside.[24]

In 1965, 90 percent of middle schools and primary schools and 80 percent of their total enrollments were in the countryside.[24]

To strengthen the political control of agricultural education, the Ministry of Education issued a directive in March 1966 calling on all educational agencies in the country to consolidate and improve the farming-study primary schools and agricultural middle schools. Specifically the directive called on the educational agencies (1) to bring politics to the fore and implement the party's policies on education; (2) to strive to firmly uphold the method of work that combines the leadership with the masses; (3) to firmly observe the principle of running schools on the basis of industry and thrift; (4) to establish a new force of teachers who have revolutionary thinking and are willing to take part in physical labor; and (5) to have the teachers rely on the leadership of the party.[25]

In June 1966 schools throughout the PRC were closed and a thorough reorientation and restructuring of the entire educational system was initiated. During the Cultural Revolution and several years thereafter, little was known about what was happening in the classrooms. Only in the fall of 1968 did reports indicate that sizable numbers of primary and secondary schools were again operating, and it was the autumn of 1970 before appreciable numbers of colleges announced that they were enrolling new students. In the countryside primary and secondary schools were placed under the management of the poor and lower-middle-class peasants because the Communist authorities considered them "the most reliable ally of the working class." In accordance with Mao Tse-tung's instructions, the period of schooling was shortened and the curriculum was changed drastically. The instructions further intensified political indoctrination, military training, and programs for promoting class struggle at the expense of basic education. Until the end of 1975, however, the final shape of the "educational revolution" was still by no means certain, for exploration and experimentation were actively continuing at all levels.

Since its establishment in 1970, Chaoyang Agricultural College of Liaoning Province has been taken as the model because of its experiences in the revolution in education. Beginning in the spring of 1975, higher educational institutions of agriculture as well as in other fields (including universities and colleges of science and engineering, medicine, and liberal arts) throughout the country have been called upon to learn from Chaoyang. The college is run in accordance with Mao Tse-tung's principle that "students should be selected from among workers and peasants with practical experience, and they should return to production after a few years' study." It has trained more than 10,000 "new" peasants in its first six years and reportedly has won high approval from the local people, primarily because it teaches Mao Tse-tung's thought to students who after graduation return to their villages to contribute to the three great revolutionary movements: class struggle, the struggle for production, and scientific experimentation.[26]

AGRICULTURAL EXTENSION

The earliest agricultural extension work in China was done in the 1910s by a few private educational and missionary institutions, including the former University of Nanking. In 1929 the first official extension regulations were issued by the Nationalist government. This was followed by the establishment of the Central Extension Committee in Nanking and extension services in 16 provinces. However, their work was brought to a standstill by the Sino-Japanese War, which began in 1937.[27]

Agricultural extension service in the PRC did not begin until 1951, when the "agricultural technique extension stations" were established in Manchuria on an experimental basis. More stations were later set up in other areas, and at the end of 1954 their number throughout the country exceeded 4,500. In general these stations were poorly organized and equipped, and most of the workers were party cadres whose technical knowledge level was low. In 1954 only about 10 percent of the 600,000 APCs and mutual-aid teams were served by the stations.

In an effort to strengthen the agricultural extension system, the Ministry of Agriculture issued a directive in 1955 urging all provinces, special districts, and hsien to set up agricultural technique extension stations. The functions of the stations were stipulated as follows: to coordinate the production experience of local peasants; to propagate modern agricultural science and technology; to help peasants increase their production and income; to promote agricultural cooperativization; and to assist local party leaders in improving agricultural techniques and management. The directive provided that the stations generally should be located in villages and rural towns with better foundations for cooperativization and better natural conditions; that in general the staff of a station should consist of no more than five persons, including technicians and one accountant, all of whom should have strong political leadership and be familiar with agricultural management; and that the technicians should be graduates of secondary agricultural schools or progressive peasants with long experience in agricultural production and at least six months of political training.[28]

The total number of extension stations increased rapidly, from 43 in 1951 to 13,669 in 1957.[29] But their operation as a whole was not efficient. In another directive issued by the Ministry of Agriculture on December 26, 1962, the stations were instructed to make improvements in administration and personnel. They were urged particularly to popularize agricultural techniques by means of demonstration and propaganda rather than by coercion, and to take local conditions into consideration.[30]

The crop failures of 1959-61 and the departure of Russian experts in the early 1960s made it necessary in 1963 for the Communist government to adopt the policy of "combining the three," under which party cadres, the scientists, and the masses were urged to cooperate and to pool their ideas. To implement this policy, research and extension systems have been reorganized and many yang pan t'ien (demonstration farms) have been set up. Agricultural scientists have been sent to these farms to test the results of research, as well as the peasants' experience, before deciding whether or not certain agricultural techniques should be popularized in larger areas. These demonstration farms, which vary in size from a few score to several thousand mou, are described as bridges between experimentation and extension. They are designed to supplement rather than to replace the experimental farms, research bases, and extension stations.[31]

During the 1960s more than ten large demonstration farms were established in the Peking municipal area, northeast China, the Szechuan basin, the Taihu area, the Pearl River delta, Hainan Island, and northwest China; and many smaller demonstration farms were set up in other areas.[32] At the end of 1964, one-fourth of China's agricultural research workers and one-half its agricultural extension workers were stationed at the some 5,400 demonstration farms set up by various agricultural institutions in 12 provinces, autonomous regions, and municipalities. In 1965 it was planned to extend the land under the guidance of demonstration farms to 50 million mou and to allocate one-half of the personnel of agricultural scientific institutions to run demonstration farms.[33]

Many scientists doubt the wisdom of their participation in the demonstration farm work and maintain that research and experimentation should be continued in the laboratories. But the authorities regard this as a bourgeois attitude.[34] It appears that one of the principal reasons for establishing the demonstration farms is to seek to fulfill a mission that state farms are supposed to fulfill but in which they apparently have failed: to serve as models of efficient agriculture.

For the dissemination of agricultural knowledge, provinces, municipalities, and autonomous regions have been urged to compile and publish local experiences in agricultural production, to hold agricultural exhibitions, to give awards to model peasants, to arrange for exchange of information between the cooperatives (now the communes), and to encourage peasants and cadres to study the advanced experience of other localities in farm management and techniques (Article 4, NPAD, 1956-67). A National Agricultural Exhibition Center has been constructed on the eastern outskirts of Peking, including five pavilions dealing with agriculture in general, rural sideline occupations, water conservation, agricultural mechanization, and animal husbandry. The predominant theme in all five pavilions is self-reliance.[35]

Radio broadcasts have been used by the government to disseminate agricultural knowledge as well as for political propaganda and mass indoctrination in the countryside. The NPAD provided that in the 7 or 12 years (depending on local conditions) beginning in 1956, the radio diffusion network was to be extended to virtually all rural areas and the majority of the cooperatives (now the communes) were to be enabled to receive the radio programs (Article 32).

Because of the vast mountainous areas and the shortage of electric power and communications facilities, since 1949 the radio broadcasting system in rural China has developed with emphasis on wired broadcasting. Under such a system hsien or communes transmit the programs they receive on the wireless, with the possible addition of their own programs. In contrast with the conventional broadcasting system, the wired network system has no tunable radio receivers at its termination points, thus requiring the audience to listen via an installed loudspeaker to whatever is broadcast by the control station.

Wired broadcasts do not need antennae and can have several broadcasting centers, even underground. The receiver is simple, containing mainly a loudspeaker. The system does not need a source of power and is less expensive than the wireless. Also, in time of war the wired broadcasts prevent the enemy from damaging the stations or jamming the broadcasts. But the rural wired system also has disadvantages. As the People's Daily stated, "In a number of communes and production brigades there is nobody to repair the loudspeakers or the wired broadcasting stations when they go wrong. People sometimes have to walk for two or three days on mountainous roads to the centers of the hsien to get repairs done. Moreover, the sound is weak in some places when rain is falling."[36]

In October 1974 Peking announced that nearly all hsien and municipalities in the country had established their own wired broadcasting stations, an "overwhelming majority" of communes had amplifier systems, more than 90 percent of the production brigades and production teams had loudspeaker systems, and 63 percent of rural households in the country had loudspeakers.[37] In September 1975, 92.7 percent of the production brigades had amplifier systems and 70 percent of the farm households had loudspeakers. There were 106 million loudspeakers in the rural areas, a 12.5-fold increase over 1965. This was said to have basically reached the targets stipulated in the NPAD.[38] However, a People's Daily article stated: "Although China has a fairly good foundation in building wired broadcasting networks, it is still unable to meet the requirements of the Party and the people."[39] The rapid expansion of the wired networks was attributed to the increased number of medium-size and small hydroelectric power stations, which provided the needed power sources in remote areas, and the development (beginning in 1969) of the carrier-current transmission system, which permitted the use of power and telephone lines for their normal functions and, simultaneously, for wired broadcasting. Development of the carrier-current system basically eliminated the need for additional poles and wires.

Although the development of wired broadcasting in China's rural areas has been considerable, the country can produce only simple loudspeakers in great numbers and will not, for some time, be able to produce radios in similar quantities. In the early 1960s transistors were becoming more common in China, and production began at Shanghai in 1963.[40] China's output of transistor radios in 1974 was reported to have registered an 11-fold increase over that in 1965, as a result of the rapid development of the country's electronics industry since 1966.[41] Transistors, however, will not take the place of the rural loudspeakers for many years because the peasants cannot afford to buy them, even if they are offered at very low prices.

In August 1973 Ta Kung Pao (Hong Kong) reported that 26 provinces and municipalities were producing television sets and that experiments with color television were being made.[42] In 1974 there were nearly 50 television stations serving the country and fewer than 200,000 sets, one for about every 4,000 people. One problem in promoting television was undoubtedly the relatively high price, which in 1974 was $250 for a black-and-white 12-inch-screen receiver made in Shanghai, the only set that national priorities permitted on the market.[43]

Since literacy plays a key role in the dissemination of agricultural knowledge, the NPAD required that illiteracy among young and middle-aged people be virtually wiped out in the 12 years beginning in 1956 (Article 31). By the end of 1959, more than 110 million, or two-thirds of all young and middle-aged people in the countryside, had been "freed from the state of illiteracy," according to Vice-Premier T'an Chen-lin.[44]

Despite the reportedly impressive networks of agricultural research, education, and extension, the quality and competence vary greatly among institutions. Some have made considerable progress, but others have accomplished very little or simply have not had many activities because of the shortage of funds and trained personnel. On the whole a slow but steady improvement seems to have been made in the 1950s and 1960s. However, the progress has been far from adequate for carrying out the arduous task of agricultural development.

Among the serious drawbacks are the requirements for the selection of staff members for the research, educational, and extension services. The most important requirement is "political level." Persons coming from an acceptable class background, especially the former poor and lower-middle-class peasants, are given high preference. Next to political standing is physical fitness—the ability to work long hours at manual labor. Technical qualification is only a poor third on the list of requirements.

Another drawback in the development of agricultural research, education, and extension is the lack of a clearly articulated organizational structure in these three services, as well as the lack of coordination among them. Progress could have been speeded up if more effective administration and management had been provided.

COLLABORATION WITH OTHER COUNTRIES

The PRC considers itself "a developing socialist country" and also a Third World country. Since 1949 Peking has established relations of economic and technical cooperation with more than 50 countries. Technical assistance in agriculture has constituted a predominant part of the agreements reached, although the number of projects in other fields, including metallurgical complexes, oil refineries, heavy machinery plants, power stations, railroads, and highways, also has gradually increased in recent years. Up to the end of 1974, some 1,000 Chinese technicians in rice, cotton, sugarcane, tea, tobacco, vegetables, tropical crops, poultry, sericulture, veterinary medicine, and other agricultural sciences had been sent to the Afro-Asian and Latin American countries.[45]

The PRC's foreign aid program was based on the eight principles, announced by Chou En-lai while he was visiting Africa in 1963-64:

1. The Chinese Government always bases itself on the principle of equality and mutual benefit in providing aid to other countries. It never regards such aid as a kind of unilateral aim but as something mutual.
2. In providing aid to other countries, the Chinese Government strictly respects the sovereignty of the recipient countries, and never attaches any conditions or asks for any privileges.
3. China provides economic aid in the form of interest-free or low-interest loans and extends the time limit for the repayment when necessary so as to lighten the burden of the recipient countries as far as possible.
4. In providing aid to other countries, the purpose of the Chinese Government is not to make the recipient countries dependent on China but to help them embark step by step on the road of self-reliance and independent economic development.
5. The Chinese Government tries its best to help the recipient countries build projects which require less investment while yielding quicker results, so that the recipient governments may increase their income and accumulate capital.

6. The Chinese Government provides the best-quality equipment and material of its own manufacture at international market prices. If the equipment and material provided by the Chinese Government are not up to the agreed specifications and quality, the Chinese Government undertakes to replace them.
7. In giving any particular technical assistance, the Government will see to it that the personnel of the recipient country fully master such technique.
8. The experts dispatched by China to help in construction in the recipient countries will have the same standard of living as the experts of the recipient country. The Chinese experts are not allowed to make any special demands or enjoy any special amenities.[46]

As in other fields, the influence of the USSR on agricultural techniques in the PRC was substantial during the Economic Recovery Period and the First Five-Year Plan. In addition to the Sino-Soviet Protocol for Scientific and Technical Cooperation signed in 1954, an agreement for technical cooperation was reached in 1957 by the academies of agricultural science in the two countries. Russian experts have assisted in planning and carrying out many agricultural projects, and Chinese students and special missions have been sent to the USSR to study Russian agricultural techniques. The USSR also has supplied the PRC with seeds of improved strains, farm machinery, and equipment.[47]

The departure of Russian experts, beginning in the 1960s was a severe blow to the technical transformation of agriculture in the PRC. The progress of many important projects was affected. For example, work at the Sanmen Gorge project, scheduled to be completed in 1961, was never finished, partly because of the withdrawal of the Russians. The construction of some of the large chemical fertilizer plants slowed down for the same reason. The supply of Russian equipment and scientific research materials also declined drastically.

Even before the Sino-Soviet split, the relationship between scientists of the two countries was not always harmonious. Many Chinese scientists frequently expressed their regret for the party cadres' absolute obedience to Russian experts. During the 1956 movement of "Letting a Hundred Flowers Bloom and a Hundred Schools Contend," numerous complaints along these lines were published in newspapers and scientific journals.[48]

In the case of the Yellow River water conservation plan, some Chinese engineers considered the presence of Russian experts as harmful. The Sanmen Gorge project was suggested by the Russians as the first project for the control of the Yellow River. Many Chinese engineers expressed the opinion that the reservoir would be filled with silt.[49] Their view was supported by some foreign observers who believed that the life of the Sanmen reservoir would be a mere half-century or so unless energetic measures in agriculture, forestry, and animal husbandry, as well as in engineering, were taken to combat silting.[50] This objection was dismissed by the Russians on the ground that soil conservation work would be undertaken. Another objection—that millions of people would have to be moved—was answered by the Russians with the argument that no large reservoir can be constructed without this problem. But perhaps the most serious objection on the part of Chinese engineers was that the Russians made recommendations without proper technical preparation. In an unmistakable tone of irony, Ch'eng Hsüeh-min wrote:

> Upon their arrival at Peking, the [Russian] experts immediately plunged themselves into intensive work. They listened to oral reports and at the same time studied all the data already compiled. After one month of investigation of the situation and of the research materials, the experts recommended that the existing data were sufficient for drawing up a plan for the Yellow River. The leadership accepted the view of the experts and decided to begin the preparation of the report on the Yellow River Plan.[51]

A number of publications on Chinese agriculture have been prepared by Russian experts during or after their visits to the PRC. Ocherki prirody i pochv Kitaia [Soils and Natural Environment of China] was prepared by V. A. Kovda in collaboration with Chinese soil scientists during a four-year period. It was published in 1959 by the Publishing House of the Soviet Academy of Sciences in Moscow.[52] Another publication, by Yu. N. Kapelinskiy et al., deals with the development of the economy and foreign economic contacts of China and contains discussions on agricultural problems.[53] A third publication, Nekotorye voprosy razvitiia sel'skokhoziaistvennoi nauki v KNR [Some Questions on the Development of Agricultural Sciences in the People's Republic of China], was published at Moscow in 1959. It contains three long reports based on the authors' surveys and observations in China: "Socialist Reform and Problems Involved in the Technological Revolution of Agriculture in the Chinese People's Republic," by V. A. Zhamin and L. A. Volkova; "Progress in Plant Physiology in the Chinese People's Republic," by B. A. Rubin; and "Accomplishments in Phyto pathology and the Organization of Plant Protection in the Chinese People's Republic," by M. V. Gorlenko.[54]

Chinese translations of Russian publications played a dominant role in introducing scientific materials from abroad, especially during the Economic Recovery Period and the First Five-Year Plan. Chao Chung-chih of the Science Press reported that many people were reluctant or did not dare to translate scientific literature from capitalist countries. From January 1, 1950, to June 30, 1956, only 11 of the 330 translations published by the Chinese Academy of Sciences and the Science Press were from publications written in capitalist countries. Some publishers accepted translations of Soviet writings only. In order to avoid "errors," writings from capitalist countries were translated into Chinese from Russian translations rather than from the original language. For example, Principles of Insect Pathology by Edward A. Steinhaus, published at New York in 1949 by McGraw-Hill, was translated into Chinese from a Russian translation. The Chinese translation, published at Peking in 1959 by the Science Press, is without many original sections deleted by the Russian publisher but includes footnotes and other materials added by the Russians.[55]

Reflecting at least a temporary cooling of the relationship between scientists of the two countries, several Chinese publications bearing the title "Soviet" have been renamed. For example, Su-lien Nung-yeh K'o-hsüeh [Soviet Agricultural Science], which began publication in 1954, was renamed Nung-yeh K'o-hsüeh [Agricultural Science] in January 1959, "with a view to including new theories, techniques and accomplishments of other countries in addition to the Soviet Union," according to an editor's note.

NOTES

1. JMJP, August 22, 1965, p. 1.
2. Edgar Snow, "Talks with Chou En-lai: The Open Door," New Republic, March 27, 1971, pp. 20-23.
3. Office of Foreign Agricultural Relations, U.S. Department of Agriculture, Report of the China-United States Agricultural Mission, (Washington, D.C., 1947), pp. 70-73, 231-37.
4. Chung-kuo Nung-pao [Chinese Agriculture Bulletin], Peking, No. 6, 1957, pp. 4-6.
5. Ch'eng Chao-hsuan, "China's Great Leap Forward in Agricultural Science," K'o-hsueh Hsin-wen [Scientific News], Peking, No. 30, 1959, pp. 18-21.
6. NCNA, Nanking, August 11, 1963.
7. Kuang-ming Jih-pao [Kuang-ming Daily], Peking, March 16, 1962, p. 1; JMJP, March 22, 1962, p. 2.
8. JMJP, October 29, 1963, p. 1.
9. NCNA, Taiyuan, March 27, 1963.
10. Wan Kuo-ting, "General Preface for the Series of Selected Works on China's Agricultural Heritage," in Ch'en Tsu-kuei, ed.,

Tao: Chung-kuo Nung-hsueh I-Ch'an Hsuan-chi Chia-lei Ti-i-chung [Rice: China's Agricultural Heritage Series, No. A-1] (Peking, 1958), pp. 1-3.

11. JMJP, July 19, 1961, p. 4.

12. Office of Foreign Agricultural Relations, U.S. Department of Agriculture, op. cit., pp. 61-70, 229-30.

13. Wang Chun, "Reform of Agricultural Universities and Colleges in Communist China," Chung-kung Yeh-chiu [Studies on Chinese Communism], Taipei, Vol. IV, No. 10 (1970), pp. 100-12.

14. "Rely on Our Own Efforts to Train More Postgraduates of Better Qualifications," editorial, Kuang-ming Jih-pao, January 17, 1966, p. 1.

15. JMJP, September 1, 1955, p. 1, and September 2, 1955, p. 1.

16. Ibid., August 25, 1956, p. 1.

17. Kuang-ming Jih-pao, July 18, 1956, p. 1.

18. Ibid., October 14, 1963, p. 1; JMJP, October 22, 1964, p. 3.

19. State Statistical Bureau, Ten Great Years (English ed.; Peking, 1960), pp. 192, 194.

20. NCNA, Peking, August 23, 1961; Jen-min Shou-ts'e [People's Handbook] (Peking, 1962), pp. 313-14; Kuang-ming Jih-pao, April 2, 1963, p. 1; Ouyang Hui-lin, "A New Type of Schools with a Bright Future," JMJP, April 2, 1963, p. 2; Ch'en Kuang, "The Seventh Anniversary of the Founding of the Agricultural Middle Schools in Kiangsu," Hung Ch'i [Red Flag], Peking, No. 4, 1965, pp. 36-47.

21. Kuang-ming Jih-pao, July 17, 1961, p. 2.

22. Chiang I-cheng, "Positively and Steadily Push Forward the Revolution of Agricultural Education; Render Still Better Service to Socialist Agriculture," Kuang-ming Jih-pao, August 19, 1965, p. 2.

23. JMJP, September 28, 1965, p. 2.

24. Ho Wei, "Properly Manage Part-Farming and Part-Study Schools; Promote Revolution of Rural Education," JMJP, July 13, 1965, p. 5.

25. JMJP, March 9, 1966, p. 1.

26. Hsüeh-hsi Chao-yang Nung-hsüeh-yuan ti Hsin-hsien Ching-yen [Learn New Experiences from the Chao-yang Agricultural College] (Peking: People's Press, 1975).

27. Office of Foreign Agricultural Relations, U.S. Department of Agriculture, op. cit., pp. 76-79.

28. Ministry of Agriculture, People's Republic of China, "Directive on the Work of Agricultural Technique Extension Stations, April 1, 1955," Jen-min Shou-ts'e (Peking, 1956), p. 482; "Agricultural Technique Extension Stations Must Be Properly Operated,"

editorial, Chung-kuo Nung-pao [Chinese Agriculture Bulletin], Peking, No. 5, 1955, pp. 22-24.

29. State Statistical Bureau, op. cit., p. 136.

30. Ministry of Agriculture, People's Republic of China, "Directive on the Strengthening of Agricultural Technique Extension Stations and Improvement of Their Work, December 26, 1962," Chung-hua Jen-min Kung-ho-kuo Fa-kuei Hui-pien [Compilation of Laws and Regulations of the People's Republic of China], Volume XIII (Peking, 1964), pp. 199-202.

31. Hsi Feng-chou, "Improve the Extension Work for Agricultural Techniques," Nung-yeh Chi-shu [Agricultural Techniques], Peking, No. 1, 1964, pp. 2-5.

32. "Demonstration Farms Are the Main Centers Through Which Agricultural Science May Serve Production," editorial, JMJP, October 25, 1964, p. 1.

33. K'o-hsueh T'ung-pao [Scientia], Peking, No. 5, 1965, pp. 459-61; Chia Pei-hua, "Farm Scientists Run Rural Demonstration Fields," Peking Review, No. 27, 1965, pp. 15-17.

34. "Properly Manage Demonstration Farms and Positively Unfold the Movement of Agricultural Science Experimentation," editorial, Kuang-ming Jih-pao, March 28, 1965, p. 2.

35. NCNA-English, Peking, April 12, 1966.

36. JMJP, January 14, 1973, p. 2.

37. NCNA, Peking, October 8, 1974.

38. Ibid., September 14, 1975, and October 3, 1975.

39. Chung Kuang, "Strive to Run Wired Broadcasting Well in the Countryside," JMJP, September 15, 1975, p. 1.

40. Ta Kung Pao, Peking, June 26, 1963,

41. NCNA, Peking, January 26, 1975.

42. Ta Kung Pao, Hong Kong, April 11, 1973, p. 1: "Big Expansion of Wired Broadcasting," Current Scene, Hong Kong, Vol. IX, No. 1, 1973; "25 Years of Radio and Television," China News Analysis, Hong Kong, No. 979, November 8, 1974.

43. John Burns, "A Night of TV in China," Christian Science Monitor, January 27, 1975, p. 7.

44. T'an Chen-lin "Strive for the Fulfillment, Ahead of Schedule, of the National Program for Agricultural Development," JMJP, April 7, 1960, p. 1.

45. Domestic Service in Mandarin, Peking, October 21, 1974, printed in FBIS, Daily Report, PRC, October 23, 1974, pp. A3-A4.

46. Chin I-wu, "Mutual Support and Sincere Cooperation," JMJP, September 22, 1974, p. 3; English excerpt in NCNA-English, Peking, September 23, 1974; printed in FBIS, Daily Report, PRC, September 24, 1974, pp. A4-A8.

47. Jen-min Shou-ts'e (Peking, 1952), pp. 354-55; Hsia T'ing, "Ten Years of Soviet Unselfish Assistance to Chinese Agriculture," Nung-yeh K'o-hsueh T'ung-hsun [Agricultural Science Bulletin], Peking, No. 19, 1959, pp. 662, 674.

48. K'o-hsueh T'ung-pao, No. 8, 1956, pp. 63-71; No. 9, 1956, pp. 54-63; No. 10, 1956, pp. 80-83.

49. Li Jui, "Several Basic Problems Concerning the Multi-Purpose Plan for the Yellow River," Hsin Chien-she [New Construction], Peking, No. 11, 1955, pp. 29-32.

50. Jean Messines, "Forest Rehabilitation and Soil Conservation in China," Unasylva, Rome, Vol. XII, No. 3 (1958), pp. 103-20.

51. Ch'eng Hsueh-min, "Great Assistance of the Russian Experts to the Yellow River Plan," in Ken-chih Huang-ho Shui-hai K'ai-fa Huang-ho Shui-li [Fundamentally Curing the Yellow River's Destructiveness and Developing the Yellow River's Water Conservation] (Peking: Propaganda Section, Ministry of Water Conservation, 1955), pp. 74-84.

52. English translation by the U.S. Joint Publications Research Service, JPRS No. 5957 (Washington, D.C., October 31, 1960).

53. English translation by the U.S. Joint Publications Research Service, JPRS No. 3234 (Washington, D.C., May 23, 1960).

54. English translation by the U.S. Joint Publications Research Service, JPRS No. 9717 (Washington, D.C., July 20, 1961).

55. Chao Chung-ch'ih, "Scientific Publication Work and 'Letting a Hundred Schools Contend,'" K'o-hsueh T'ung-pao, No. 11, 1956, pp. 48-50.

CHAPTER 9
EXPANDING THE AREA UNDER CULTIVATION

Compared with raising the yield per unit of land, expansion of the area under cultivation is a longer and more costly process to increase agricultural production. Despite this fact and despite China's limited potentialities of additional land for agricultural uses, the Communist government has made strenuous efforts to bring new land under cultivation. The general situation of China's arable land and cultivated land and the progress of the land reclamation program in the 1950s and 1960s will be discussed in the sections that follow.

Until a nationwide survey is made, any enumeration of the total area of China's arable land and cultivated land must be only a rough estimation. Table 9.1, based on estimates from various sources, indicates that at present China's land suitable for farming is only about 15 percent of its total land area and that one-fourth of the land area presently classified as arable is still uncultivated.

ARABLE LAND

China is a rugged country for the most part, a land of mountains. Level land is found along some river valleys, in delta regions, and in a few other areas. The largest of the level areas is the compound delta that reaches from northern Chekiang Province northward into Manchuria. It includes the north China plain, which is the most important region in the production of wheat and cotton. There are basins around lakes Poyang and Tungting and a few alluvial areas south of the Yangtze valley, including small deltas in the vicinity of Swatow and Canton. West China has limited level areas in the alluvial fan of the Chengtu plain in the Szechuan and the plateau of Yunnan. In Inner Mongolia,

TABLE 9.1

Land Use in the PRC

	Thousand Mou	Percentage of Total
Arable land	2,200,000	15.28
Cultivated	1,650,000	11.46
Uncultivated	550,000	3.82
Forest area	1,140,000	7.92
Land suitable for afforestation	4,500,000	31.25
Grassland	4,000,000	27.78
Other (deserts, etc.)	2,560,000	17.77
Total	14,400,000	100.00

Sources: Chao Shih-ying, "Searching the Direction of China's Agriculture," K'o-hsueh T'ung-pao [Science Bulletin], Peking, No. 8, 1963, pp. 1-5; Food and Agriculture Organization of the United Nations, FAO Production Yearbook, 1969 (Rome, 1970), p. 6; Chao Hsi, "Our Country's Agricultural Future Is Infinitely Good," K'o-hsueh Ta-chung [Popular Science], Peking, No. 5, 1963, pp. 1-3; Sung Chia-t'ai, "Change the Appearance of Our Country's Geography by Using Land Rationally," Ti-li Chih-shih [Geographical Knowledge], Peking, No. 4, 1956, pp. 144-49; Chia Sheng-hsiu, "Grassland," K'o-hsueh Ta-chung, No. 9, 1964, pp. 323-24.

Sinkiang, and Tibet, there are some areas of level or rolling country, mostly in the form of high plateaus. Otherwise, the country is mountainous.

Table 9.2 shows China's land area classified according to elevation by Wong Wen-hao, a well-known Chinese geologist. Areas below 500 meters are mostly suitable for farming. Areas between 500 and 1,000 meters are for the most part hilly, but agriculture can be operated profitably with the help of terracing and other water and soil conservation devices. Areas at 1,000-2,000 meters' elevation contain mostly plateaus where animal husbandry is more profitable than farming. The bulk of the remaining land consists of mountainous areas with very little agricultural productivity.

Thus, from the topographical viewpoint, only 14 percent of China's land is suitable for farming. Even with the possible addition of 18 percent of land at 500-1,000 meters' elevation, the total area suitable for farming is still less than one-third of China's total land area. Moreover, a substantial portion of the land under 1,000 meters' elevation is located in areas where climatic and soil conditions are not favorable for farming because of such factors as uneven distribution of precipitation throughout the year, short growing season, and

TABLE 9.2

Elevation of China's Land Areas

Elevation (meters)	Area (square kilometers)	Percentage of Total Land Area
0-500	1,344,000	14
500-1,000	1,728,000	18
1,000-2,000	3,360,000	35
2,000-3,000	768,000	8
3,000-4,000	480,000	5
4,000-5,000	384,000	4
Over 5,000	1,536,000	16
Total	9,600,000	100

Source: Chang Lun-p'o, Chung-kuo ti Tzu-yuan [China's Resources] (Peiping, 1947), p. 1. Based on estimates made by Wong Wen-hao.

poor soils. Another problem is that most of the arable land considered suitable for expansion is in Manchuria and Sinkiang, where a number of obstacles must be overcome before any large-scale reclamation program can be carried out effectively. Most of the areas lie on the frontiers, away from trade routes and good harbors. In order for the land to be usable, agricultural development must be accompanied by transportation, housing, public health facilities, local industry, and electric power.

CULTIVATED LAND

The total area of a country's cultivated land, like that of arable land, can hardly remain static for any length of time. It fluctuates in response to the interplay of many factors, including land reclamation policy, success or failure of harvests in recent years, rates of rent, political stability or instability, migration program, and changes of agricultural techniques.

In 1947 China's cultivated land was estimated by the Nationalist government at 1,410,731,000 mou (235,122,00 acres), about 10 percent of the total land area of the country.[1] Official data reported in Table 9.3 indicated that China's cultivated land increased slowly but steadily from 1,468,220,000 mou (244.7 million acres) in 1949 to 1,677,450,000 mou (279,575,000 acres) in 1957, or from 10.20 percent to 11.65 percent of the total land area (14.4 billion mou or 2.4 billion acres). Thus, the average annual increase from 1949 to 1957 was less than 0.2 percent. The slow increase was confirmed by Vice-Premier Teng Tzu-hui, who

TABLE 9.3

Cultivated Land in China, 1949-58

Year	Cultivated Land (thousand mou)	Percentage of Total Land Area
1949	1,468,220	10.20
1950	1,505,340	10.45
1951	1,555,070	10.80
1952	1,618,780	11.24
1953	1,627,930	11.31
1954	1,640,320	11.39
1955	1,652,350	11.47
1956	1,677,370	11.64
1957	1,677,450	11.65
1958	1,616,800	11.23

Note: Percentages are based on the estimated total land area of 14.4 billion mou.

Source: State Statistical Bureau, Ten Great Years (English ed.; Peking, 1960), p. 128.

stated at the National Conference on the Work of State Farms and State Pastures held in 1957: "For the past few years, the total area of our country's cultivated land increased less than 1 percent annually, and yet her annual population increase was about 2.5 percent."[2]

In 1958 the total area of cultivated land decreased to 1,616,800,000 mou (269,467,000 acres), or 11.23 percent of the total land area. The decline was not explained; but it could have been caused, at least in part, by the short-lived campaign to "sow a smaller acreage, realize a higher yield, and reap a greater harvest," launched late in 1958. Under the "three-thirds system" the communes were asked to use one-third of the existing farmland for growing principal crops, one-third to lie fallow or for growing fodder crops and green manure, and one-third for reservoirs or for growing trees and ornamental plants. However, several months later a policy promoting the opposite extreme was adopted, and the communes were urged to "sow a larger acreage and reap a greater harvest."[3]

Official data on the total area of cultivated land have not been released since 1958. On the basis of estimates made by various sources and figures used in the Communist Chinese press, it would appear that throughout the 1960s or so, China's total area of cultivated land hovered around 1.65 billion mou (275 million acres). or 11.5 percent of the country's total land area. Thus, China's cultivated land is less than two-thirds that of the United States, although China has a population more than three times as large as that of the United States.[4]

TABLE 9.4

Low-yielding Farmland in the PRC, 1958

	Cultivated Land (thousand mou)	Low-Yielding Farmland (thousand mou)	Low-Yielding Farmland as a Percentage of Cultivated Land
Peking	2,000	200	10.0
Tientsin	1,402	300	21.4
Liaoning	72,390	18,750	25.9
Honan	134,616	35,000	26.0
Shensi	67,181	17,870	26.6
Shantung	138,000	40,000	29.0
Hopei	133,400	40,300	30.2
Hupeh	64,725	20,000	30.9
Yunnan	42,368	13,600	32.1
Kiangsu	94,273	30,450	32.3
Kiangsi	42,424	14,000	33.0
Anhwei	89,021	30,000	33.7
Kirin	70,716	24,680	34.9
Sinkiang	27,933	10,000	35.8
Hunan	58,300	21,570	37.0
Kweichou	30,763	11,600	37.7
Chekiang	33,766	13,000	38.5
Heilungkiang	110,204	43,200	39.2
Chinghai	7,576	3,000	39.6
Shanghai	514	210	40.8
Inner Mongolia	83,398	34,110	40.9
Kansu	72,639	30,000	41.3
Szechuan	115,418	51,130	44.3
Shansi	66,208	31,250	47.2
Fukien	22,363	10,600	47.4
Kwangsi	38,000	19,000	50.0
Kwangtung	59,048	31,000	52.5
Total	1,678,686	594,820	35.4

Source: Chung-kuo Nung-pao [Chinese Agricultural Bulletin], Peking, No. 8, 1958, p. 25; No. 10, 1958, p. 20.

Moreover, much of the cultivated land in China is of poor quality. According to the National Soil Survey conducted in 1958, 400 million mou, or nearly one-fourth of the 1.65 billion mou of cultivated land in the PRC, needed immediate improvement; 1,200 million mou, or three-fourths of the total, needed crop rotation; and 200 million mou, or one-eighth of the total, needed grading to make irrigation and drainage more effective. Also, the fragmentation of much of the farmland in the south had to be readjusted to facilitate cultivation.[5]

It was also reported in 1958 that more than one-third of the farmland in the PRC was low-yielding—saline, alkaline, acidic, sandy, swampy, or erosive (see Table 9.4). Low-yielding farmland constituted large portions of the cultivated land in virtually all the provinces and autonomous regions and in the three largest municipalities—Shanghai, Peking, and Tientsin. The municipality of Peking had the lowest ratio, 10 percent. Figures for other political divisions ranged from 21.4 percent for Tientsin to 52.5 percent for Kwangtung.[6] In 1962 there were about 300 million mou of alkaline and saline soils in northwest, northeast, and north China and the coastal areas of northern Kiangsu. Of this total nearly one-third, or 100 million mou, were in the cultivated land in north China and 20 percent of its irrigated area.[7]

RECLAMATION OF WASTELAND

Of China's some 1.5 billion mou (250 million acres) of wasteland, about 800 million mou had been preliminarily surveyed by the end of 1957; 500 million mou could be reclaimed "at a comparatively low capital cost," according to Chiang Ch'i-hsien, vice-minister of state farms and land reclamation. These 500 million mou are presumably the land generally referred to as arable but uncultivated.[8]

Major portions of the wasteland already surveyed are in Manchuria; the Sinkiang-Uigur Autonomous Region; Chinghai; the coridor west of the Ordos bend of the Yellow River in Kansu; the tropical and subtropical areas of Fukien, Kwangtung, Kwangsi, and Yunnan; the areas near the Gulf of Po Hai along the Liaoning, Hopei, and Shantung seacoasts; and some areas in Szechuan, Hupeh, and northern Kiangsu.[9]

The NPAD provided that in the 12 years beginning in 1956, the area reclaimed by state farms should be increased from the 1955 figure of over 13 million mou (2.2 million acres) to about 100 million mou (16.7 million acres). The program also urged the APCs (now the communes) to make full use of scattered bits of idle land, such as ridges and corners of land plots, the edges of ponds, land along ditches, and unused threshing grounds. Factories; mines; commercial and agricultural enterprises; departments of education, culture, public health, water conservation, communications, and military affairs; and mass organizations also were asked to exercise the strictest economy in using land for capital construction and particularly to refrain from using cultivated land for such purpose (Article 16).

In addition to the goals for land reclamation prescribed in the NPAD, 1956-67, the First Five-Year Plan (1953-57) also called for the expansion of 38.68 million mou of cultivated land, mainly through reclamation of small pieces of wasteland by organized peasants, so that at the end of 1957 the total area of cultivated land would be increased to 1.65 billion mou, as compared with 1.61 billion mou in 1952.[10] The increase actually accomplished in the First Five-Year Plan period was reported to be 58.67 million mou, bring the total area of cultivated land at the end of 1957 to 1.67 billion mou.[11]

It was planned to reclaim 50 million mou during the Second Five-Year Plan period (1958-62).[12] According to a report by Vice-Premier Po I-p'o, for 1958 alone the goal was 17,735,000 mou, of which 7 million mou were to be reclaimed by the state farms under the Ministry of Land Reclamation and 10.73 million mou by the state farms under local governments and by peasants and imigrants.[13] The total acreage of wasteland actually reclaimed during the First Five-Year Plan period and the acreage expected to be reclaimed during the Second Five-Year Plan period was thus 108,670,000 mou. However, another source reported that during the years 1949-60, only 70 million mou (11.7 million acres), or less than 5 percent of the estimated total area of wasteland (1.5 billion mou), were reclaimed.[14]

Data on the progress of land reclamation are fragmentary and, in many cases, conflicting. Following are brief accounts of the accomplishments and problems in Manchuria, Sinkiang, and other areas, as given in various publications and press reports.

Northeast China (Manchuria)

The northeast China (Manchurian) plain consists of the flat valleys of the Nonni, Sungari, and Liao rivers with a total area of about 460 million mou, which is comparable in size to the north China plain and larger than the lower Yangtze valley. A large part of the plain, especially in Heilungkiang Province, has rich virgin soil. The rolling country is suitable for mechanized farming. Although the climate here is cold and the frost-free period is short, the hot summer has long hours of sunshine. The rainfall in the area is more abundant than in north China, ranging from 20 to more than 30 inches per year.

Manchuria has about 230 million mou of arable wasteland already surveyed.[15] Of this total, 100 million mou are in Heilungkiang Province, where reclamation work began even before the Communist takeover. During the blockade of the Sino-Japanese War, a brigade of the Communists' Eighth Route Army was sent to Nanniwan, near Yenan in

Shensi Province, to reclaim wasteland for food production. In 1946 the Chinese Communists established Hua Ch'uan Water Conservation Farm on the grassland in the part of Heilungkiang Province that is commonly known as the "Northern Great Wilderness." Several collective farm groups were later set up in the area.[16]

In 1964 there were more than 400 state farms of various sizes in Manchuria, cultivating a total of some 15 million mou.[17] Many of the state farms are located in the Sanchiang [Three-River] Plain Reclamation Area, with the Heilungkiang River in the north, the Ussuri River in the east, and the Sungari River running through the area. The plain occupies about 24 million mou, of which 15 million are arable. One of the largest state farms in the Sanchiang Plain Reclamation Area is the much-publicized Friendship State Farm, with 492,000 mou of land; it was established in 1954 with technical aid and a supply of tractors, harvesters, threshers, mechanized five-share plows, trucks, and other equipment from the USSR.[18]

In 1965 there were 36 state farms in the Sanchiang Plain Reclamation Area, equipped with more than 6,500 tractors, 1,500 combine-harvesters, and a network of farm machinery repair plants. In 1964 the area produced nearly 1,000 million chin of grains and soybeans and delivered more than 500 million chin of grains to the state, the largest quantity ever delivered.[19]

But, according to a Soviet technical team that participated in setting up the Friendship State Farm, although most of the Sanchiang area is suitable for farming, there are certain unfavorable climatic, soil, and hydrological conditions in the area that must be overcome in order to obtain higher and more stable yields of the principal crops.[20] One of the main obstacles to the reclamation of this area is waterlogging in the summer, which forms marshes. In 1957 and 1958, for instance, waterlogging brought heavy losses to the Friendship State Farm, and its sown acreage and grain production decreased considerably.[21]

Sinkiang

Although the Sinkiang-Uighur Autonomous Region occupies about one-sixth of China's total land area, its dry climate, rough mountains, and extensive deserts make agriculture profitable only in small areas along the foot of the Kunlun and Tienshan mountains. The region is the driest part of China, with an average annual rainfall of five to six inches. Deserts occupy about one-third of the land. The rate of evaporation under the scorching sun is very high, and irrigation is indispensable to agriculture. For centuries farming in this region has depended on water from melted snow and glaciers from the Tienshan Mountains, which flows through a few inland rivers, and on numerous irrigation canals and ditches. But most of these waterways have become seriously silted. Melted snow rushes down the mountains in summer, often causing floods. In the spring and autumn, when water is most needed, periodic drought threatens.[22]

Sinkiang has some 100 million mou (16.5 million acres) of wasteland with soil sufficiently fertile to support agriculture, according to a survey conducted in the 1950s by the Chinese Academy of Sciences. Since 1956 the academy has sent groups of scientists to Sinkiang to study the problems of developing wasteland for agricultural uses, such as the possibility of utilizing the water from glaciers and snow for farming in a section of the Tienshan range near Urumchi; measures for protecting farmland against sandstorms in the desert areas of southwestern and northern Sinkiang; conditions of the grasslands in the livestock-breeding hsien of T'oli in northern Sinkiang; and methods of improving alkaline soil in other parts of the autonomous regions.[23]

An important portion of the land reclaimed in Sinkiang is under the administration of the Sinkiang Production-Construction Corps (PCC) of the PLA. In 1961 the PCC administered more than 18.9 million mou, or one-third of the cultivated land in Sinkiang.[24] Of Sinkiang's estimated 1968 population of 8 million, about 1.2 million, or one-seventh, were PCC workers and members of their families.[25]

The PCC concept stemmed from the Yenan period prior to World War II, during which the Red Army cooperated closely with civilians in agricultural and other production work. In 1953, the first PCC unit was formally installed in Sinkiang and was assigned to carry out programs for industrial and agricultural development, including the operation of state farms. Thus, while retaining its status as a national defense unit of the PLA stationed in Sinkiang, the PCC also plays an important role in economic reconstruction under the supervision of relevant agencies of the State Council and the Sinkiang Committee of the CCP. Since the eruption of the Sino-Soviet clash at Chenpo Island in 1969, the PCC has been introduced into the other frontier regions, including Inner Mongolia, Heilungkiang, Chinghai, and Tibet.

Chinese youngsters have been drafted in increasing numbers from all parts of the country to join the PCC for hard labor. By May 1965 their number exceeded 260,000, including 50,000 from Shanghai. The PCC serves as an organization for settling boys and girls, mostly from middle (secondary) schools, who are recalcitrants or who will not go further in their studies. A land reclamation college for the training of agricultural personnel in Sinkiang was established at Tarim in 1958 by the PCC. The college combines education with productive labor through the part-time work and part-time study system. In 1965 it had 670 students and a teaching staff of 90. By the end of that year, it had graduated 950 students at intermediate and higher levels in the fields of plant science, agricultural mechanization, water conservation, animal husbandry, agricultural economics, and forestry. Students at the college are mostly children of lower-echelon cadres, farmers, and state farm and reclamation area employees.[26]

Since adequate irrigation is prerequisite to the expansion of cultivated land in Sinkiang, numerous canals, ditches, and reservoirs have been constructed by the communes and state farms to improve the water supply of the region as well as to harness the Tarim, the Manass, the Urumchi, and other principal rivers.[27] In recent years emphasis has been given to the digging of wells in order to utilize more of Sinkiang's underground water. The communes and the state farms are reported to have sunk thousands of wells, many of which are artesian wells with a depth of up to 100 meters.[28]

As a result of improved water conservation and increasing mechanization, Sinkiang's total area of cultivated land was reported to have doubled from 18.11 million mou (3,018,000 acres) in 1949 to 36.17 million mou (6,028,000 acres) in 1964.[29] By the end of 1963, the Sinkiang PCC had established more than 140 mechanized state farms and had reclaimed some 11 million Mou (1.83 million acres) of wasteland.[30]

The major portions of wasteland reclaimed in Sinkiang during the 1950s and 1960s included (1) areas totaling more than 1.5 million mou (250,000 acres) along the Tarim, Sinkiang's principal river, producing grains and oil-bearing, industrial, and fodder crops; (2) the Urumchi River basin, an important supplier of grains, vegetables, and meat for Sinkiang's capital city of Urumchi; (3) the Khotan (Hotien) area, a narrow greenbelt of oases extending 600 kilometers from east to west and fringing the southern part of the Taklamakan Desert, producing wheat, cotton, and oil-bearing crops; (4) some 850,000 mou of wasteland in the Masowan area reclaimed for growing wheat, cotton, and corn; (5) the Kawako reclamation area of some 16,000 mou established in 1960 within the Taklamakan Desert, producing grains, cotton, and oil-bearing crops; (6) areas at the foot of the Tienshan, Kunlum, and Altai mountains; and (7) the Ili and Manass valleys.

Land reclamation and increased yield per unit of land reportedly have enabled Sinkiang to produce three times as much grain and six times as much cotton in 1965 as in 1949 and to be self-sufficient in both.[31] The region has been developed into an important producer of long-fiber cotton so that some of the cotton fields in other provinces can be used for growing grains and other principal crops. Cotton production has been particularly successful in the Turfan basin, the Manass River valley, Ku-erh-le, and Agsu because these areas have plenty of sunshine and irrigation facilities are available. The average annual increase of cotton output during the First Five-Year Plan period was 16.54 percent, and the average yield of 410,000 mou of cotton farms operated by the Sinkiang PCC in 1957 was 68.8 chin per mou, which was about double the yield of many cotton farms in the interior. But the costs of reclaiming land for growing cotton in Sinkiang—including the costs of leveling, soil improvement, farm machinery, draft animals, and construction of canals, ditches, roads, and houses—were much higher than in the interior.[32]

One source estimated that the total area suitable for growing cotton in Sinkiang may reach 23 million mou, mostly in the southern part of the region. On the basis of an average yield of 70 chin per mou, the annual output would be 16 million tan.[33] But a number of problems must be solved before cotton production in Sinkiang can be substantially increased. These include fertilization, cultivation technique, and control of diseases and insect pests, especially Agrotis segtis Hubner, which is the insect most harmful to cotton in Sinkiang.[34]

In the livestock industry, as a result of artificial insemination and control of animal diseases, Sinkiang has become a supply center of fine breeds of sheep and horses for the entire country. In 1966 Siakiang had the second largest number of sheep, ranking after Inner Mongolia. But it ranked first in the production of fine and semifine wool. The autonomous region has developed the gungnais sheep, the first fine-wool breed of sheep in China, and the famous Ili horse. The gungnais sheep is known for its large size, delicious mutton, adaptability to different natural conditions, and high propagation rate. The fineness, color, and length of the wool are suitable for making high-grade fabric. The Ili horse can cover 30 to 40 kilometers a day for three or four days in a row and can draw a load of 1,500 kilograms.[35]

Despite the reported accomplishments the reclamation work in Sinkiang has confronted extreme hardship, especially along the Tarim, the principal river of the region, with a total length of 2,750 kilometers. The Tarim's major tributaries, including the Khotan, Yarkand, Kashgar, and Agsu rivers, have the greatest flow in the late spring and early summer, when they are fed by melting snow and glaciers from the northern slopes of the Tienshan mountains. The water hurtles down the mountain slopes in heavy torrents, causing serious floods in some areas. In the dry season a large part of the river water evaporates or disappears in the desert sand before reaching the Tarim, causing serious droughts in many places. Moreover, because of the rapid evaporation of underground water, land easily becomes alkaline. Various water and soil conservation measures have been taken, but land conditions have not improved very much. A similar situation exists in other reclaimed areas in Sinkiang.[36]

From the viewpoint of agriculture, Sinkiang is not an asset to China. Land reclamation in the region has not been successful, even at high cost, because of the unfavorable natural and economic conditions. But the PRC government apparently is building up the region at all costs because of its strategic importance vis-à-vis the Soviet

Union, as well as the minor nationalities and anti-Communist or anti-Mao elements.

Other Areas

In addition to the land reclamation projects in Manchuria and Sinkiang, many have been carried out or planned in other provinces and autonomous regions. The more important projects include the following:

1. Some 90,000 mou of sandy wasteland near the Ulanpuho Desert in eastern Inner Mongolia.[37]

2. Some 300,000 mou of wasteland in the Karmu area south of the Tsaidam basin in Chinghai.[38]

3. Areas near the Maowusu Desert bordering northern Shensi, Inner Mongolia, and Ninghsia.[39]

4. Areas on the edge of the Scrub Desert, locally known as "the Hunger Desert," in Ninghsia.[40]

5. Some 900,000 mou of saline wasteland along the Yellow River in Ninghsia.[41]

6. A narrow strip of land extending 2,400 shih li (750 miles) from east to west in the Kansu corridor, which was China's ancient route to the West.[42]

7. Some 19,500 mou of marshy, alkaline wasteland, formerly the abandoned course of the Wei River, a tributary of the Yellow River, in Shensi.[43]

8. Vast stretches of saline wasteland along the north China coast have been transformed into rice fields. This is the continuation of a project started by a Japanese reclamation company to produce rice for Japanese troops during World War II. Included in the project was the reclamation of more than 500,000 mou of saline wasteland on the northern Shantung Peninsula along the shores of Po Hai Bay, from which the sea receded centuries ago.[44]

9. The Yellow River delta, known among local peasants as the "Isolated Island" and formerly a desolate delta at the estuary of the Yellow River in Shantung. The entire area was almost untouched before 1950.[45]

10. About 1.25 million mou along lakes Tungting and Poyang, two of China's largest freshwater lakes.[46]

11. More than 2.25 million mou of seacoast and lakeside wasteland in Kiangsu.[47]

12. Nearly 130,000 mou of land along Hangchou Bay at the mouth of the Chientang River.[48] Hangchou Bay is a major cotton- and hemp-growing area in China.

13. Some areas on Hainan Island and the Luichow Peninsula in Kwangtung for growing rubber, spices, coconuts, coffee, and other tropical and subtropical crops.[49]

Exactly how much wasteland has been reclaimed by the state farms is not known. Many of the figures released are conflicting and confusing. In some cases, for example, it is not clear whether a figure refers to the acreage of land reclaimed in a specific year or to the total acreage so far reclaimed. In other cases it is not clear whether a figure refers to the land reclaimed by the state farms only or to that reclaimed by the communes as well. In light of the available data, it seems certain that the pace of land reclamation by state farms has been far behind the target prescribed in the NPAD: to increase reclaimed wasteland from 13 million mou in 1955 to 100 million mou in 1967. The slow pace is evidenced by the fact that from 1949 to 1958 cultivated land in China increased by only a little over 1 percent (see Table 9.3, above); by the admissions of several high government officials, including Chou En-lai and Teng Tzu-hui; and by the reticence about the progress of land reclamation since the Leap Forward period.[50]

There have been numerous claims about the yields of the principal crops produced by state farms. In 1952 the average yield of wheat produced by the mechanized state farms was reported to be 21.7 percent higher than that of wheat produced by all farms in the PRC; the average yield of cotton, 48.7 percent higher; and the average yield of soybeans, 11.0 percent higher. In 1956 the average yield of grain crops produced by the state farms was reported to be 4 percent higher than that of grain crops produced by the APCs; and the average yield of cotton, 66 percent higher.[51] In 1957 Vice-Premier Teng Tzu-hui revealed that the average yield of grains produced by the state farms in that year was 220 chin per mou; that of soybeans, 120 chin per mou; and that of cotton, 57 chin per mou.[52] The average yield of grains for all farms in the country in that year was 204 chin per mou; that of soybeans, 105.1 chin per mou; and that of cotton, 37.9 chin per mou.[53]

Despite these claims it has been reported that the labor efficiency of the state farms in general is low. Many of them spend extravagantly and have suffered a deficit.[54] Also, the use of farm machinery and the methods of cultivation and management need improvement, according to Wang Chen, minister of state farms and land reclamation.[55] At a national conference on state farms and land reclamation held at Peking in March 1966, some achievements of state farms were praised; but it was pointed out that for many of them politics had not been brought to the fore, management and operations needed adjustment, production level was not high enough, capital construction on farmland was not solid, and farming skills were on a "relatively crude level."[56]

Simultaneously with the land reclamation by the state farms, the APCs (now the communes), especially those with surplus labor, have been urged to cultivate as much of the wasteland as possible, in addition to the existing farmland, in their own or neighboring areas. It was pointed out by the People's Daily that although a major portion of China's wasteland is located in the northeast, northwest, and other border areas, there are innumerable small pieces of wasteland, totaling some 100 million mou (16.7 million acres), that are scattered throughout the country, including wasteland along rivers, canals, and seacoasts and around buildings. Most of these small fragments of land are of low productivity, but if cultivated, they may produce 20 billion chin (13.1 million tons) of grains each year on the basis of 200 chin per mou (1,600 pounds per acre), according to the People's Daily.[57] Very few data are available on the land reclaimed by the APCs (communes). The total acreage of such land probably is considerably smaller than that reclaimed by the state farms.

In evaluating the PRC's land reclamation program, it is important to note that substantial portions of the land reclaimed in the 1950s and 1960s are located in sparsely populated upland areas, on the edge of deserts ravaged by shifting sands, on arid grasslands marked by drastic changes in climate, or in areas handicapped by alkaline and saline soil. Most of these lands are agriculturally unproductive.

China's deserts total approximately 1.6 billion mou (265 million acres), an area nearly equivalent to the total area of existing cultivated land, and are distributed mainly in the Sinkiang-Uighur, Inner Mongolia, and Ninghsia-Hui autonomous regions and in Chinghai, Kansu, and Shensi provinces. The principal deserts include the Taklamakan Desert (the Big Gobi), China's largest desert, with a total area of 450 million mou (75 million acres); the Kurbantongut Desert in Sinkiang; and the Tengri, Patachilian, and Ulanpuho deserts in Inner Mongolia. The Communist government believes that nearly 300 million mou (50 million acres) of the deserts can be improved to grow some plants or developed into livestock bases.[58]

As an initial step toward desert transformation, a surveying team composed of members of the Chinese Academy of Sciences and Soviet scientists was organized in 1957 to study deserts in Inner Mongolia, the Yülin Special District in northern Shensi, the Ninghsia plain, and the corridor west of the Yellow River. Among the principal subjects studied were the natural conditions of these areas and farmers' experience in controlling the sand that might be adopted for improving the deserts.[59]

The schemes used for desert transformation in the PRC include the construction of irrigation and drainage systems, the extensive cultivation of grass and green manure crops, the leveling of sand dunes, and the planting of tree shelter belts to ward off windstorms. Among the many desert control projects the conquest of shifting sand to enable the Paotow-Lanchow Railroad to cross the southeastern end of the Tengri Desert in Ninghsia is considered by Communist authorities to be one of the most successful. This railroad, China's first to traverse a desert, is reported to have remained open to traffic since its completion in 1958 and, according to a 1961 report, service was not interrupted by the desert after the spring of 1959.[60] The train passes at six places through the Tengri Desert, altogether a distance of 40 kilometers.

The method used for the project, as described by the New China News Agency, was to divide the ground along the railroad into small rectangles and surround each rectangle with half-buried wheat stalks to prevent the sand from being blown away. The checkered sandbreaks were then covered with gravel, clay, and asphalt and planted with grass, in order to slow wind velocity be some 20 percent and greatly reduce sand movement within the sandbreaks. Among the 30 kinds of plants that were tried out in 1956-64, shrubs of clover, yellow willow, and oleaster grew well on the sand dunes, which vary in height from dozens of meters to more than 100 meters. In the sandbreaks where plants have grown, thin solidified crusts have appeared on the surface of the sand. In the opinion of Chinese scientists, this change in the formation of the soil is the result of fixing the sand and the growth of plants. They believe that if such crusts were general, the sand would no longer move.[61]

Despite many claims it was admitted that desert transformation remains a long and arduous task and that many problems are yet to be solved. In the first place, the methods and measures for the control of deserts have not been very efficient and the cost has been prohibitively high. Second, many deserts can be transformed only through the utilization of underground water, which often contains too much salt, making it difficult to develop agriculture, forestry, and animal husbandry. Finally, the development of deserts calls for large quantities of motive power. Although there is plenty of wind and solar energy in the deserts, how to make effective use of such energy remains to be studied.[62]

As already stated, mountains and hills constitute a major portion of China's land area. The Communist government has made efforts to cultivate some mountains and hills for production of grains as well as forest products, fruits, and other crops. The NPAD urged the APCs (now the communes) to develop grain production in some of the mountainous areas, provided this would not affect the water and soil conservation program; to grow timber forests in the remote mountainous areas; and to grow forests for firewood and charcoal as well as fruits in the less remote mountainous areas. In the south emphasis has been given to the cultivation of tea-oil and tung-oil trees, bamboo, mulberries, and tea plants; in the north to that of walnuts, Mongolian oaks, and wild pepper. The cooperatives were also encouraged to grow medicinal herbs in the mountainous areas (Article 17).[63]

Party cadres, young intellectuals, and others have been sent to the mountainous areas in increasing numbers, particularly during the Leap Forward campaign, to participate in such activities as reclaiming waste mountain valleys. In south China the cultivation of fruit trees on hilly land has been especially encouraged in order to save as much flat land as possible for planting grains and cotton.[64] It has been reported that some 40 million young intellectuals were sent to the mountainous areas for rural work during 1955-65.[65] Several million more were drafted during the Cultural Revolution for similar purposes.

Despite the reported progress the surge into the mountainous areas was greeted with little enthusiasm by the cadres and peasants. The Communist press observed in 1958 that "conservative thinking and inertia among some of the cadres should be subjected to excoriating criticism."[66] The situation apparently remained the same in more recent years. In a 1966 editorial the Liberation Daily stated that not enough attention had been paid to the development of mountainous areas, owing to erroneous thinking on the part of some cadres, who maintained that the development of such areas is not worth the trouble. Other cadres were afraid of hardship in mountainous areas and reluctant to go there.[67]

In addition to deserts and mountains, efforts have been made to improve China's vast area of grassland for developing animal husbandry. China's grassland, with an estimated total area of 4 billion mou (667 million acres), or nearly 30 percent of the country's total land area, is mainly distributed west of the diagonal extending from the Greater Khingan Mountains southwestward toward the eastern foreland of the Tibet Plateau. It constitutes a major portion of Sinkiang, Chinghai, Inner Mongolia, Ninghsia, and the Kantze and Ahpo districts in Szechuan.

Flat grassland makes up approximately two-thirds of the total area of China's grassland, while the remaining one-third consists mainly of hilly grassland. Most of the country's grassland lies in arid zones marked by drastic changes in climate. The grass yield is unstable and varies a great deal between years and seasons. Moreover, much of the grassland in China is suitable only for grazing and not for harvesting grass. For this reason the stock of hay is inadequate to meet the needs of the animal herds. Mismanagement and excessive grazing have worsened the situation. Another problem in many areas is the inadequate supply of drinking water for men and animals. In some of these areas, the underground water resources are rather abundant, but their utilization has not been well developed. Rodent and insect pests also constitute a serious menace to the grassland. In Chinghai particularly, the damage to grassland by rodents has been very serious; and the grassland in Chinghai and Sinkiang frequently has been attacked by locusts.

In recent years measures reportedly have been taken for the improvement of grassland, with special attention given to its better cultivation and utilization. Reservoirs, wells, ditches, dams, and other water conservation projects have been carried out to store water and channel the runoff from accumulated snow. Scientific research has been planned on methods to raise production on the grassland and the establishment of grassland experimental stations.[68]

In the vast Gobi area in Bayan Naoh League of the Inner Mongolian Autonomous Region, rich underground water resources were reported to have been discovered 20 to 50 meters below the surface in 1959 by a desert control team from the Chinese Academy of Sciences. It was believed that at least one-third of the 180,000 square kilometers of the Gobi could be brought into use for grazing if water were provided. As a result of improved water supply, Bayan Naoh League's livestock increased from 1.3 million head in 1949 to 3.8 million head in 1963.[69]

Part of the 8,000-square-kilometer Tamchin grassland, one of the largest arid grasslands in Inner Mongolia, was improved in 1963 through the sinking of large numbers of artesian and pump-operated wells and the construction of aqueducts, canals, and reservoirs. Formerly known as "the demon's domain" for its desolation, the Tamchin grassland rises like a flat, green terrace 100 meters above the surrounding pastures in the central part of Inner Mongolia. It does not have any river, and shortage of drinking water for men and animals was the main obstacle to using the rich grassland. Now it can feed 100,000 head of livestock every year, according to press reports reports.[70]

A question has often been posed as to whether some of China's grassland should be improved for animal husbandry or transformed into forests. China is short of both livestock and forest products, and the problem of soil erosion is serious. From the standpoint of ecological conditions at least, it would seem that some of the grassland, especially the hilly grassland, is rather unpromising for the development of animal husbandry. Therefore, the only use to be gained from it would be through afforestation.[71]

Since 1949 reclamation projects have been carried out on some of the alkaline and saline land with reportedly successful results. From the viewpoint of land utilization, it seems more economical and profitable for China to reclaim some of its land on the long coast than to reclaim the deserts, mountains, and grassland. Some of the low-lying marshy areas along the seacoast that are now unproductive because of poor drainage and harmful accumulation of alkali salt can be reclaimed. In southeast China particularly, precipitation is sufficient to wash the salt off the reclaimed coast and the growing season for crops is long. Some alkaline land can also be reclaimed in the semiarid parts of north, northeast, and northwest China, where water is available and topography is favorable. Before World War II considerable areas of saline land along the coast in northern Kiangsu were successfully drained and irrigated for the growing of cotton; and during the Japanese occupation thousands of acres in the vicinity of the Tientsin-Tangku area in Hopei and in several areas of northeast China that formerly were considered worthless were reclaimed for growing rice.[72]

ACQUISITION OF FARMLAND FOR CONSTRUCTION PURPOSES

As population grows, such nonagricultural land requirements as housing, government buildings, industry, highways, airports, and cemeteries steadily encroach on the arable land area. The procedure for the requisition of land for such purposes was first provided in the "Regulations Governing Agrarian Reform in Suburban Areas," approved by the State Administration Council (now the State Council) in November 1950, and again the "Regulations Governing the Requisition of Land for State Construction," promulgated in December 1953. As the construction program proceeded, it became more and more apparent that these regulations were far from adequate.

The municipality of Peking can serve as an example. Up to the end of June 1957, a total of 210,000 mou of land in Peking's suburban areas had been requisitioned for building construction. Of this total, 15,000 mou was wasteland and the remaining 195,000 mou was farmland.[73] The arbitrary manner of the requisition can be seen in many of the notices, almost identical in form, issued by the Land Administration Bureau or the City Planning Commission, which appeared frequently in the newspapers. The one in the People's Daily of March 29, 1952, reads:

> It has now been decided to start building construction in the following areas: [description of location and boundaries]. All land and graveyard owners in the area prescribed above are expected to settle the transfer of land, the removal of graves and other matters with their respective village governments within fifteen days from the date at which this notice first appears in the newspaper. If any owner fails to do so, the grave will be treated as ownerless and will be removed by the government to the People's Cemetery.

In the pre-Communist days it was the custom in China for each family to have its own burial grounds. As a rule the geomancer located the grave site. Since he gave attention only to the feng shui (the supposed magical wind and water influences), the grave was as often located in the center of a cultivated field as elsewhere. This practice created a real problem not only because of the land actually taken from cultivation but also because of the obstacle the grave presented to cultivation. J. Lossing Buck estimated in 1937 that if the graves could be removed, China's crop area for the eight agricultural areas, which totaled 232 million acres, would be increased by 1.1 percent or 2,552,000 acres, and that this increased area could support over 400,000 additional farm families.[74]

Cremation and cemeteries have now replaced the private burial system, but many of the lands and graveyards requisitioned during the 1950s and 1960s were left unused. Moreover, in many construction projects too much space was allowed between buildings, causing even greater waste. It was estimated in 1958 that if the space between buildings had been planned more properly, 4,000 or 5,000 mou of the arable land requisitioned in Peking's suburban areas by 33 government units and other institutions could have been saved, to yield about one million chin of grain per year.[75]

In view of the waste, in 1957, 88 government units and institutions were asked to return land that they had requisitioned but on which construction projects had not been started. At the end of May 1957, a total of more than 7,800 mou of such land was reported to have been turned over either to the APCs for cultivation or to the government for reallocation. However, some pieces had to remain idle because they had been either enclosed by buildings or divided into too many fragmentary parts.[76]

The 1953 regulations were revised in January 1958. Under the revised regulations the area of land to be requisitioned must be approved by "the organ which is competent to approve the preliminary design of the construction project concerned"; in cases where the removal of more than 30 households was involved or more than 300 mou of land was called for, the allotment must be approved by the people's council at the provincial level; and a clear-cut system was provided for constant supervision and inspection of the use of land requisitioned.

Despite these revised regulations, the land acquisition situation apparently has not improved very much. On June 6, 1962, the People's Daily carried a number of letters from readers that disclosed that in recent years some capital construction units had taken excessive quantities of land. Much of it was unused for a long time or was used by government organs and factories for cultivating subsidiary food crops, with unsatisfactory harvests. The newspaper again urged the respective units to return the land to the production teams for cultivation, "in accordance with the successive instructions on the economical use of land."[77]

In another article, the authors stated:

> The economic level of our country is still not very high and it is impossible for the State to devote too much financial, material and labor resources to the reclamation of wasteland. Moreover, newly-reclaimed wasteland will yield a relatively low output in the first few years, and the use of one mou of cultivated land for construction purposes must be offset by the reclamation of several mou of wasteland. Thus, it is clear that we should pay full attention to the economical use of cultivated land for construction projects.[78]

PROSPECTS FOR EXPANDING THE AREA UNDER CULTIVATION

From the above discussion it is apparent that the attempts by the Communist Chinese regime to expand cultivated areas in the 1950s and 1960s have not been successful and that the prospects for the future are not bright. In the first place, there is very limited acreage of uncultivated land suitable for agricultural use. In fact, a sizable portion of the present cultivated land is not profitable for agricultural use and should be developed for other purposes, such as forestry, animal husbandry, and recreation. Also, a substantial portion of the land presently classified as arable is marginal or submarginal land with steep slopes, little or unreliable precipitation, and other unfavorable natural conditions; and it is possible that such land should not be considered arable at all. It is true that some of the land considered not arable at present may become suitable for farming through the introduction of modern farm machinery and other technological innovation, but the acreage of such land is negligible and will remain so for some time.

Furthermore, the greater part of China's wasteland is situated in remote frontier regions. Tremendous financial and technical problems are involved in any large-scale reclamation program. Surveying work, power machinery, better transportation facilities, and well-planned migration are needed. These, in turn, would require abundant investment of capital, manpower, equipment, and material, as admitted by Chou En-lai and other Communist Chinese leaders.[79]

Among Chinese scientists the question of whether China's cultivated land could be extended by reclaiming wasteland has been discussed frequently. Even those who think that this could be done raised serious problems. Huang Ping-wei, director of the Research Institute of Geography, Chinese Academy of Sciences, and his associate, Shih Ch'ien, are of the opinion that the present 1.65 billion mou (275 million acres) of cultivated land could be doubled. However, Huang maintains that most of the land reclaimed would be mountainous and erosive and that there would be too much salt in the soil. He suggests that this could be ignored, possibly for three to five years, but in eight or ten years' time serious damage would occur. Such damage might involve not only the newly reclaimed land but also neighboring land, and even land at a considerable distance.[80] Shih Ch'ien expressed the opinion that owing to the shortage of machinery, reclamation of wasteland on a large scale is not possible.[81]

The zeal to exploit wasteland has produced unfavorable results. The dikes at the sides of the Grand Canal have been destroyed in many places because trees have been cut down so that wheat can be planted.[82] Several readers wrote to the People's Daily in 1962, complaining that the cultivation of steep land in certain areas had caused soil erosion. The "Provisional Regulations Governing the Conservation of Water and Soil," issued by the State Council in 1957, specified that reclamation of wasteland on slopes having a gradient of 25 degrees or more generally should be prohibited. It also provided that the gradient of a slope on which cultivation is prohibited may be determined variably in accordance with the condition of soil, the intensity and volume of rainfall, the distribution of forests, farmland, and population, and other conditions. Apparently the regulations were ignored during the zeal to exploit wasteland.[83] For example, a survey of northern Shensi and western Shansi in 1963 indicated that much of the reclaimed land was on slopes with a gradient of 25 degrees or more, and in some cases the gradient exceeded 60 degrees. The reclamation was carried out by individual production teams without coordination with water and soil conservation work. As a result forest land and grassland were destroyed.[84] In many areas the exploitation of wasteland has led to the permanent destruction of valuable resources in exchange for a relatively short-term gain in agricultural output.

Recognizing the difficulties of land reclamation, in recent years the Communist government has given increasing emphasis to the development of "stable and high-yielding farmland despite drought and waterlogging," farmland that guarantees high yields even in the years of natural calamities. This is not an easy task. The People's Daily noted:

> The area of farmland which is stable and high-yielding despite drought and waterlogging at present constitutes only a small part of the 1.65 billion mou of all farmland. In the near future, say in three to five years, it will still be only a small part. The development of stable and high-yielding farmland not only takes a long time but also

requires a huge force of manpower and large amounts of capital and materials. In certain areas, it requires the construction of some large and medium-sized projects for water conservation, afforestation, soil improvement and other purposes. It would be difficult for the communes and localities to initiate such large projects, although there are certain small projects in which they could actively participate, such as the digging of irrigation channels in the fields, leveling of land and heavier application of fertilizers.[85]

NOTES

1. Directorate of Statistics, National Government of China, Chung-hua Min-kuo T'ung-chi T'i-yao [Statistical Abstracts of the Republic of China] (Nanking, 1947), pp. 1, 14.

2. Teng Tzu-hui, "Great Accomplishments of State Farms and State Pastures in the Past Few Years and Their Basic Functions in the Future," Chung-kuo Nung-pao [Chinese Agricultural Bulletin], Peking, No. 8, 1957, pp. 2-6.

3. "Resolution on Certain Questions Concerning the People's Communes," Hsin-hua Pan-yueh-k'an [New China Semimonthly], Peking, No. 24, 1958, pp. 3-11; JMJP, editorial, June 11, 1959, p. 1.

4. Food and Agriculture Organization of the United Nations, FAO Production Yearbook, 1969 (Rome, 1970), pp. 4-6.

5. Ministry of Agriculture, People's Republic of China, "Notice on the Strengthening of the Planning Works in Land Utilization of the People's Communes," Chung-kuo Nung-pao, No. 18, 1959, p. 18.

6. Chung-kuo Nung-pao, No. 8, 1958, p. 25; and No. 10, 1958, p. 20.

7. Chang Tzu-lin and Huang Jung-han, "The Problem of Improving Alkali and Saline Land and Preventing Alkalization and Salinization in North China," Hung Ch'i [Red Flag], Peking, Nos. 15/16, 1962, pp. 47-56.

8. Chiang Ch'i-hsien, "Strive to Fulfill the Reclamation Work of 100 Million Mou of Wasteland Ahead of Schedule and to Manage Well the State Farms," Chung-kuo Nung-k'en [Chinese Agricultural Reclamation], Peking, No. 1, 1958, pp. 4-5.

9. Teng Tzu-hui, "Great Accomplishments of the State Farms and Pastures in the Past Few Years and Their Basic Functions in the Future," Chi-hsieh-hua Nung-yeh [Mechanized Agriculture], Peking, No. 4, 1957, pp. 1-5; Kuang-ming Jih-pao [Kuang-ming Daily], Peking, October 9, 1961, p. 1; Liu Huan-wen, "Miscellaneous Sketches of Hohsi," Kuang-ming Jih-pao, July 20, 1965, p. 4.

10. Chung-hua Jen-min Kung-ho-kuo Fa-chan Kuo-min Ching-chi Ti-i-ko Wu-nien Chi-hua 1953-1957 [The First Five-Year Plan for the Development of the Economy of the Chinese People's Republic, 1953-1957] (Peking, 1955), p. 85.

11. Hsin-hua Pan-yueh-k'an, No. 8, 1959, p. 50.

12. Chiang Ch'i-hsien, op. cit.

13. Po I-p'o, "Draft Resolution on the National Economy Plan for 1958, "a report at the Fifth Session of the First National People's Congress, February 3, 1958, Hsin-hua Pan-yueh-k'an, No. 5, 1958, pp. 12-23.

14. Kuang-ming Jih-pao, October 9, 1961, p. 1.

15. Teng Tzu-hui, op. cit.

16. Lu Jen, Hsin-chung-kuo Ti-i-ko Chi-t'i Nung-chuang ti Ku-shih [The Story of New China's First Collective Farm Group] (Hankow, 1952), pp. 1-7.

17. NCNA-English, Shenyang, February 17, 1964.

18. Ho Hsu-chih, "The Site of the Friendship State Farm-the "Three-River Plain' in Northeast China," Ti-li Chih-shih [Geographical Knowledge], Peking, No. 4, 1955, p. 111; Kuo Wen-yu, Kuo-ying Nung-ch'ang [State Farms] (Peking, 1957).

19. China News Service, Canton, news release No. 4146, August 9, 1965, and No. 4174, September 6, 1965; NCNA-English, Shenyang, February 17, 1964; NCNA-English, Harbin, January 13, 1966.

20. Chang Yu-san and Wang Pao-kung, "Land Planning of the Friendship State Farm," Chi-hsieh-hua Nung-yeh, Shenyang, No. 10, 1955, pp. 18-24.

21. T'ien Lien-ch'ien and Sun Kuang-hsien, "Friendship State Farm on Its Way to Healthy Development," JMJP, May 3, 1962, p. 2; "Diversified Undertakings Centered on One Single Enterprise," editorial, JMJP, May 3, 1962, p. 1.

22. Wang Wei-p'ing and Hu Ting-mei, Hsin-chiang Wei-wu-erh Tzu-chih-chu [The Sinkiang-Uighur Autonomous Region] (Peking, 1959), pp. 14-24.

23. NCNA, Urumchi, May 11, 1961, and May 27, 1963.

24. Min-tsu T'uan-chieh [Unity of Nationalities], Peking, No. 12, 1961.

25. NCNA, Urumchi, September 6, 1968.

26. Kuang-ming Jih-pao, December 29, 1964, p. 2, and March 15, 1966, p. 1; "Tarim Institute of Land Reclamation—A New Type of School in Sinkiang," Peking Review, No. 41, 1966, pp. 27-30.

27. NCNA-English, Urumchi, September 16, 1965.

28. Ibid., February 24, 1966.

29. Sai Fu-ting, "Ten Years of Great Achievements in Sinkiang's Agriculture," Chung-kuo Nung-pao, No. 19, 1959, pp. 19-23; NCNA-English, Urumchi, November 13, 1964.

30. JMJP, October 30, 1963, p. 1.

31. NCNA-English, Urumchi, March 19, 1966.

32. Kuo Ch'un-hua, "The Potentiality of Reducing the Cost for Constructing State Farms in Sinkiang," Chung-kuo Nung-k'en [China Agricultural Reclamation], Peking, No. 1, 1958, pp. 4-5.

33. Ts'ao Tse-yuan, "The Outlook of Land Reclamation for Growing Cotton in Sinkiang," ibid., No. 5, 1957, pp. 25, 29-30.

34. "Several Problems of Cotton Production in Sinkiang," Chung-kuo Nung-k'en, No. 5, 1958, pp. 14-15.

35. Ta Kung Pao, Peking, March 31, 1966, p. 2; NCNA-English, Urumchi, November 26, 1970.

36. Kung-jen Jih-pao [Daily Worker], Peking, September 27, 1965, p. 1.

37. NCNA-English, Huhehot, June 16, 1963.

38. Ibid., Sining, April 26, 1966.

39. Ibid., Sian, September 7, 1963.

40. Ibid., Yinchuan, February 27, 1964.

41. Ibid., May 2, 1966.

42. Ibid., Lanchou, January 3, 1966.

43. Ibid., Sian, September 5, 1961.

44. Ibid., Tientsin, September 20, 1961; Tsinan, May 22, 1963.

45. Ibid., Tsinan, October 17, 1961.

46. Ibid., Peking, December 9, 1960; Nanchang, January 6, 1964.

47. Ibid., Nanking, October 18, 1961.

48. Ibid., Peking, July 20, 1970.

49. Lin Hsin, "Achievements and Future Plans of the State Farms in Kwangtung," Jo-tai Tso-wu [Tropical Crops], Canton, No. 15, 1959, pp. 1-4.

50. Planning Commission, Government of India, Report of the Indian Delegation to China on Agrarian Cooperatives (New Delhi, May 1957), p. 11; Teng Tzu-hui, op. cit.

51. Wang Chen, "Strengthening the Construction of State Farms," Hung Ch'i, No. 7, 1961, pp. 1-7.

52. Teng Tzu-hui, op. cit.

53. "Data for the National Agricultural Exhibition," Nung-yeh K'o-hsueh T'ung-hsun [Agricultural Science Bulletin], Peking, No. 2 1959, pp. 46-47.

54. JMJP, January 9, 1965, p. 2.

55. Wang Chen, op. cit.

56. JMJP, April 1, 1966, p. 2.

57. Ibid., January 16, 1958, p. 1.

58. Li Ch'iao, "China's Deserts," K'o-hsueh Ta-chung [Popular Science], Peking, No. 6, 1963, pp. 6-8; Desert Laboratory, Lanchou Research Institution of Glaciers, Frozen Grounds, and Deserts,

Chinese Academy of Sciences, Chung-kuo Sha-mo Kai-lun [A Brief Discussion of China's Deserts] (Peking, 1974).

59. Tuan Shao-po, "Transforming Deserts into Good Farms and Gobi into Grassland," K'o-hsueh Hau-pao [Science Pictorial], Peking, No. 9, 1959, pp. 330-31.

60. JMJP, September 28, 1961, p. 1.

61. NCNA, Yinchuan, October 6, 1963.

62. Lou T'ung-mao, "A Brief Account of the Deserts in Northwest China," Ti-li Chih-shih, No. 9. 1959, pp. 413-14; Li Ch'iao, op. cit.

63. Wen Ping-ya, "Developing a Mountain Area," China Reconstructs, Vol. XXIV, No. 11, November 1975, pp. 35-57.

64. Li Lai-jung et al., Nan-fang ti Kuo-shu Shang-shan [Cultivating Fruit Trees on Hilly Land in South China] (Peking, 1956), pp. 1-120.

65. JMJP, December 9, 1965, p. 1.

66. Jen-min Shou-ts'e [People's Handbook] (Peking, 1958), pp. 528-32.

67. "A Great Deal Can Be Done in Tapping the Potentialities of Mountainous Areas," editorial, Chieh-fang Jih-pao [Liberation Daily], March 4, 1966, reprinted in JMJP, March 15, 1966, p. 2.

68. Wang Chi-wu, The Forests of China (Cambridge, Mass., 1961), pp. 171-76; Chia Shen-hsiu, "Problems on the Utilization and Improvement of Grassland in China," Kuang-ming Jih-pao, February 23, 1964, p. 4.

69. NCNA-English, Huhehot, April 10, 1963; and November 27, 1964.

70. Ibid., July 10, 1963.

71. Jean Messines, "Forest Rehabilitation and Soil Conservation in China," Unasylva, Rome, Vol. XII, No. 3 (1958), pp. 103-20.

72. Office of Foreign Agricultural Relations, U.S. Department of Agriculture, Report of the China-United States Agricultural Mission (Washington, D.C., 1947), p. 4; Shen Tsung-han, Agricultural Resources of China (Ithaca, N.Y.: Cornell University Press, 1951), pp. 84-86.

73. JMJP, November 27, 1957, p. 1.

74. J. Lossing Buck, Land Utilization in China (Nanking: University of Nanking, 1937), pp. 178-79.

75. JMJP, January 7, 1958, p. 1.

76. NCNA, Peking, June 11, 1957.

77. "A Problem Which Should Be Speedily Solved," editorial, JMJP, June 6, 1962, p. 1.

78. Miao Hui-fen and Chang Ch'eng-p'o, "On the Economical Use of Land for Construction Projects," JMJP, March 5, 1964, p. 5.

79. Chung Chieh, "A Discussion on the Question of Agricultural Modernization in Our Country," Ching-chi Yen-chiu [Economic Research], Peking, No. 12, 1963, pp. 73-78; Planning Commission, Government of India, loc. cit.

80. Huang Ping-wei, "Several Problems to Be Considered in Expanding the Acreage of Cultivated Land," JMJP, May 5, 1961, p. 7.

81. Kuang-ming Jih-pao, October 9, 1961, p. 1.

82. JMJP, November 13, 1961, p. 2.

83. Ibid., March 23, 1962, p. 2.

84. Chao Ming-fu, "Water and Soil Conservation in Regions Along the Middle Reaches of the Yellow River," JMJP, November 14, 1963, p. 2.

85. "Properly Manage All Farmland with a View to Building Stable and High-Yielding Farmland," editorial, JMJP, July 4, 1964, p. 1.

CHAPTER

10

RAISING THE YIELD PER UNIT OF LAND

Because of the limited potentialities for expanding land under cultivation, most plans for increasing agricultural production have been geared to improving the yield per unit of land. The NPAD called for higher yields of grains, cotton, and other principal crops and laid down the targets as follows:

1. During the 12 years beginning in 1956, the average annual yield of grains should be raised from the 1955 figure of 150 chin per mou to 400 chin per mou (from 1.12 to 3.00 metric tons per hectare) in the areas north of the Yellow River, the Chinling Mountains, the Pailung River, and the Yellow River in Chinghai Province. In the areas south of the Yellow River and north of the Huai River, the yield should be raised from the 1955 figure of 208 chin per mou to 500 chin per mou (from 1.55 to 3.73 metric tons per hectare). In the areas south of the Huai River, the Chinling Mountains, and the Pailung River, the yield should be increased from the 1955 figure of 400 chin per mou to 800 chin per mou (from 3.00 to 6.00 metric tons per hectare).

2. In the same 12 years, the average annual yield of ginned cotton should be raised from the 1955 figure of 35 chin per mou to 40, 60, 80, or 100 chin per mou (0.30, 0.45, 0.60, or 0.75 metric tons per hectare), depending on local conditions.

3. While giving priority to increasing grain production, various localities were expected to exert efforts to develop a diversified agriculture to meet the state production targets for textile raw materials, oil-bearing crops, sugarcane, sugar beets, tea, cured tobacco, fruits, vegetables, medicinal herbs, tropical and subtropical crops, and other economic crops (Article 2).

The approximate boundaries of the three principal grain-growing regions, commonly referred to by the Chinese Communist authorities as the "400-chin region," the "500-chin region," and the "800-chin

MAP 10.1

The Three Grain-Producing Regions of China

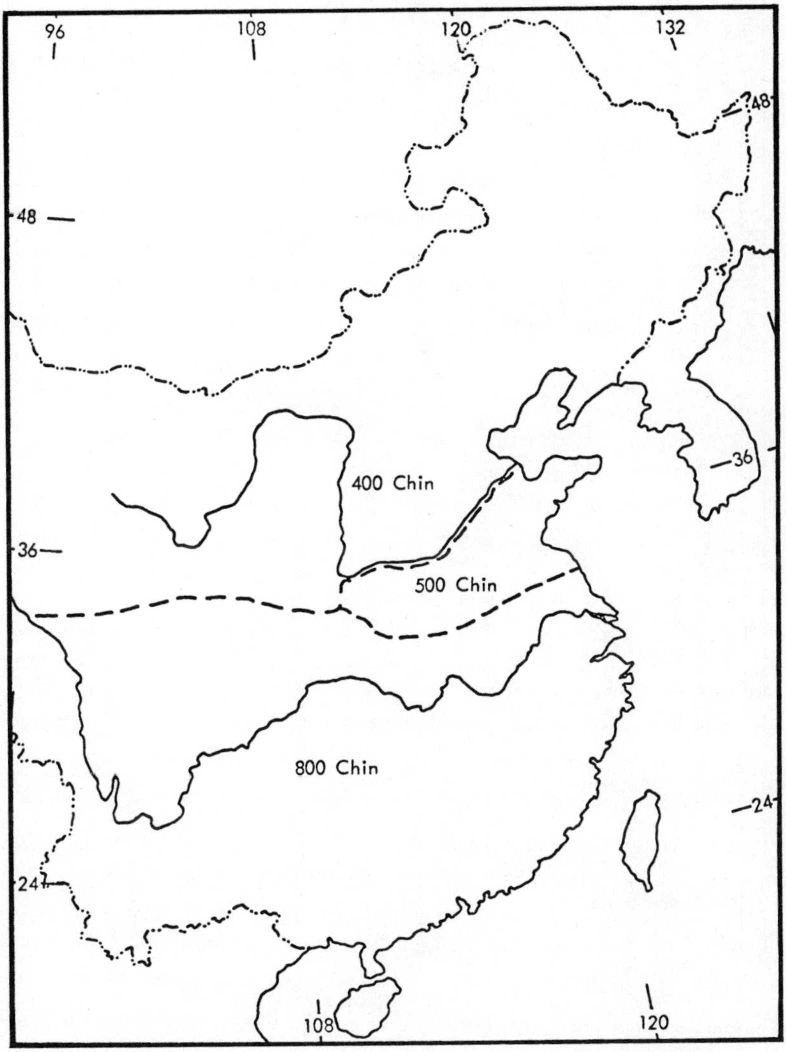

Source: Compiled by the author.

TABLE 10.1

Natural Conditions of the Three Regions Specified in the NPAD, 1956-67

	400-Chin Region	500-Chin Region	800-Chin Region
Location	Areas north of the Yellow River, the Chinling Mountains, and the Pailung River	Areas south of the Yellow River and north of the Huai River	Areas south of the Huai River, the Chinling Mountains, and the Pailung River
Acreage of cultivated land	730 million mou	290 million mou	620 million mou
Annual precipitation	600 mm. in the northeast; 630 mm. in eastern Hopei; less than 200 mm. in Sinkiang; 400 mm. in other areas	Average 600 mm.	1 billion mou in the Yangtze River basin; 1.5 - 2 billion mou in the areas further south.
Growing season	120-170 days north of the Great Wall; 200 days in the Liaotung Peninsula area and south of the Great Wall	220 days or longer	250-300 days in the Yangtze River basin; practically frost-free in the areas further south
Principal crops	Wheat, corn, millet, kaoliang, soybeans, cotton, flax, sugar beets	Food crops (about the same as those in the 400-chin region), cotton, tobacco	Rice, corn, potatoes, wheat, ramie, jute, sugarcane, cotton, tobacco, rapeseeds, tea, silk cocoons, rubber, coffee, citronella, tropical fruits
Number of crops per year	Generally one crop per year north of the Great Wall; three crops in two years in the Liaotung Peninsula area and south of the Great Wall	Three crops in two years in most areas; five crops in three years or two crops per year in some areas	Two or three crops per year in the Yangtze River basin; three crops per year in the areas further south

Source: Hsin-hua Pan-yueh-k'an [New China Semimonthly], Peking, No. 2, 1958, p. 94.

region," are shown in Map 10.1. Both the boundaries of these regions and their respective grain production targets appear to have been set arbitrarily, although their natural conditions, as shown in Table 10.1, are apparently different.

To attain the goals cited above, the NPAD, provided various technical measures for raising yields per unit on the existing cultivated land. Eight of these measures were considered by Mao Tse-tung as the most important and were referred to as the "Eight-Character Charter of Agriculture." The eight Chinese characters were shui (water conservation), fei (fertilization), t'u (soil conservation), chung (plant breeding), mi (close planting), pao (plant protection), kung (tool improvement), and kuan (field management). Although these eight measures were selected on a rather arbitrary basis and do not necessarily represent all the important phases of agricultural technology, the "Eight-Character Charter of Agriculture" includes the areas of concentration in planning and carrying out the program of technical transformation of agriculture since the Leap Forward campaign. For this reason the problems of raising the yields per unit of the existing cultivated land will be discussed in terms of these eight aspects in the following eight chapters.

CHAPTER

11

WATER CONSERVATION

Because of the poorly distributed rainfall through the seasons and the frequent occurrence of floods and droughts, water is regarded by Mao Tse-tung as the lifeline of China's agriculture and is placed above all other factors in his "Eight-Character Charter of Agriculture." The NPAD provided that in the 12 years beginning in 1956, ordinary floods and droughts should in the main be eliminated (Article 5).

During the earlier years of the regime, many large water conservation projects were carried out with great enthusiasm and fanfare. It soon became evident that many of the projects did not perform their functions as satisfactorily as expected, because of improper planning, lack of coordination, and poor construction. The authorities also realized that large projects would require large amounts of funds, raw materials, and labor, and would take a long time to complete.

Therefore, in mapping out the NPAD emphasis was given to small and medium-size water conservation projects, such as digging of wells and ponds; construction of dikes, ditches, reservoirs, and embankments; and harnessing of small rivers (Article 5). This policy of emphasizing small and medium-size projects was reaffirmed at the beginning of the Leap Forward campaign in 1958, when a "three-priority" program of water conservation was adopted. Under the program priority was given to the following: (1) small projects, supplemented by large and medium-size projects, in order to gradually form coordinated networks; (2) water-storage projects, although considerable attention was also to be paid to drainage projects, in order to prevent land from being waterlogged, salinized, and alkalized; and (3) projects undertaken by the communes under the unified planning and help of the state. Only large projects were to be built by the central government.[1] This remains the guiding principle for water conservation work at present.

Especially since the Cultural Revolution, commune members have been asked to participate in construction of water conservation projects during the slack season after autumn harvest. In Hunan Province more than 6 million people, or approximately one-sixth of the population, reportedly took part in water conservation work during the winter of 1971-72. There were 210,000 projects of various sizes under construction. The participants received no wages for their labor, which, however, was recorded in work points.[2]

According to Peking Review, water conservation work completed in the PRC since 1949 included 130,000 kilometers of dikes repaired, reinforced or built; nearly 100 big canals for draining floodwater and excess water; more than 2,000 large and medium-size reservoirs; 1.3 million power-operated wells (which irrigate 7.3 million hectares of land), and mechanized irrigation and drainage facilities with a total capacity of more than 30 million horsepower.[3]

For the convenience of discussion, the progress of the water conservation program in the PRC may be reviewed in terms of two major aspects: flood control, particularly the harnessing of the principal rivers, and drought control, particularly irrigation. The two are of course interrelated, and many water conservation projects have been constructed or planned for the control of both floods and droughts.

FLOOD CONTROL

Approximately one-fourth of the cultivated land in the PRC is subject to floods and waterlogging. This land is distributed primarily in the areas along the middle and lower reaches of the Yangtze River; the Huai River basin in northern Kiangsu and northern Anhwei; the plains along the lower reaches of the Yellow River in Shantung and Honan and along the Hai River in Hopei; the Sungari-Liao plains in northeast China; and the Pearl River delta. These are among the most populated and most fertile areas in China, and they are also the principal areas producing grains and cotton. The Yellow, Yangtze, Huai, and Hai rivers are the most treacherous rivers in China, although there is also frequent flood menace along other principal rivers.

The Yellow River

There have been nine major course changes in the history of the Yellow River, which for centuries has been known to Westerners as "China's sorrow." Until 1194 the river reached the sea at various points on the Po Hai north of the Shantung Peninsula. In that year one arm entered the sea south of Shantung through the valley of the Huai

River. After nearly seven centuries of southeasterly flow, the Yellow River resumed its northeasterly course about 1855 in what is essentially its present channel. The river was again diverted toward the Huai River basin in June 1938, but this time its dike was broken intentionally at a point northwest of Kaifeng by the Nationalist government to slow the Japanese advance. The breach was sealed in March 1947 with the assistance of the United Nations Relief and Rehabilitation Administration (UNRRA), and the Yellow River returned to its former northeasterly course.

The Communist regime has given renewed attention to the problem of flood control along the Yellow River. The program includes a complex system of flood-detention reservoirs in the upper reaches, the strengthening of dikes along the lower course, and the provision of spillways, particularly near the mouth, where ice jams frequently cause flooding in the spring. A multiple-purpose plan for the development of the Yellow River basin, mapped out with Soviet assistance, was adopted in 1955. The plan included the construction of reservoirs for irrigation and flood control, the development of hydroelectric power for industries, and the extension of the navigable sections of the Yellow River. One of the first projects under the plan was the much publicized Sanmen Gorge project above Loyang in Honan Province, which began in April 1957. It was expected that when completed the project could reduce the flow of the river from 37,000 cubic meters per second to 8,000 cubic meters per second. According to a Chinese Communist account, the project included the construction of a dam 110 meters high and 960 meters long, to create a reservoir capable of holding 35 billion cubic meters of water and second in size only to the largest reservoir in the world, the Kuibyshev Reservoir in the Soviet Union.[4]

In 1959 it was reported that the Sanmen Gorge project would be completed by October 1960, two years ahead of schedule; no further information about its progress was revealed, however, until early 1975, when Peking disclosed that the project, originally designed "by foreign experts," had been rebuilt. A major reconstruction effort, starting in 1965, was required to clear away the silt and to make the project suitable for the generation of electric power as well as for flood control.

According to the Chinese sources, the reservoir was in fact completed in September 1960 but "because of faulty estimates of the volume of silt flow, the reservoir became badly silted up some time after being put into use." The accumulation of silt, deriving from the easily eroded loess rock in the upper Yellow River basin, was so serious that the silted area spread upstream past the town of Tungkwan and into the Wei River, a tributary of the Yellow River. This not only threatened to render the reservoir useless but also endangered the safety of the city of Sian.

The reconstruction of the project proceeded in two stages. From 1965 to 1968 four of the steel pipes that were originally designed to carry water to power turbines were rebuilt to carry both water and silt out of the reservoir. Two additional silt discharge pipes were installed in the dam. From 1970 to 1973 "still other modifications and changes were made in the original design and a new water turbine generator resistant to mud, silt and friction was designed, manufactured and installed." The original power capacity of the dam had been scaled down, and it was now designed to generate 200,000 kilowatts, a fifth of the original Soviet design. The first of four 50,000-kilowatt units began generating power in December 1973. The original Soviet design was for eight 137,500-kilowatt units. The first was delivered in 1960, before the withdrawal of Soviet aid, and was reported to have been installed by 1963; no power production is believed to have been started at that time, however, apparently because of the reported silting problems in the reservoir.

As a result of the reconstruction, the Sanmen Gorge Dam is said to have the following advantages: relieving the lower reaches of the threat of major floods; lifting the threat of ice floes; providing abundant water for irrigation, urban consumption, and industrial uses; and making possible the use of the natural river flow to drive the turbines.[5]

The Yangtze River

Although the Yangtze carries less sediment than the Yellow River (600 million tons a year compared with 4 billion tons) and aggrades its channel to a lesser extent, it nevertheless presents a summer flood threat. Disastrous floods occurred in 1931 and 1954. The most serious floods of the Yangtze River usually have occurred along its middle reaches. As a first step toward regulating the Yangtze, the Ching River flood-detention project was carried out in 1952. The Ching is a section of the Yangtze between Chinkiang hsien in Hupeh Province and Chenlingchi at Lake Tungting in Hunan Province. Along the Ching River is an important area of 860 hectares producing rice and cotton. It is also one of the localities of the Yangtze River basin that has been most seriously affected by floods. Principal measures taken to control the floods since 1952 have been strengthening of main dikes, diversion of the river water, and construction of two main dams.[6]

Work was completed in early 1974 for the first stage of a key water conservation project in the gorge near Tanchiangkou on the upper reaches of the Han River, the major tributary of the Yangtze. The 1,530-kilometer-long Han River rises at the southern foothills of the Chinling Mountains in northwest China and flows southeast through 87 hsien and cities in Shensi, Honan, and Hupeh provinces to empty into the Yangtze at the city of Wuhan. The Han River basin covers an area of 174,000 square kilometers and is one of China's largest industrial and agricultural areas. The Tanchiangkou project is the key to permanently harnessing the Han River and to multipurpose exploitation of its water resources—flood prevention, power generation, irrigation, navigation, and fish breeding.[7]

For further control of the Yangtze River, a plan has been mapped out to complete a number of reservoirs and strengthen the dikes along the river's middle and lower reaches, and to build a big dam near the "Three Gorges" in Szechuan in order to slow the flow of the upper reaches. No details of the plan have been revealed. However, it has been admitted that the plan involves many financial and technical problems.[8]

The Huai River

The Huai is the largest river between the Yellow River and the Yangtze River and drains the southeastern margin of the north China plain. A relatively short river (675 miles), the Huai is subject to severe summer flooding. When the Yellow River usurped the channel of the Huai before 1855 and again from 1938 to 1947, the invading stream deposited vast amounts of sediment in the Huai channel, disrupting its natural course to the sea. The Huai discharged its water into the fluctuating lakes Hungtse and Kaoyu, which spread over vast areas in times of flood. Part of the water from the naturally mouthless Huai followed the Chungshan River and the Grand Canal to the sea.

In response to Mao Tse-tung's call for "assuring success in bringing the Huai River under control" in 1950, the Peking regime embarked on an ambitious program that included the construction of detention reservoirs (some with power-generating capacity) on the upstream tributaries, the dredging of the main channel, the strengthening of dikes, flood-diversion dams, reservoirs in middle course, and the digging of a new outlet to the Yellow Sea. A new outlet for the Huai, the North Kiangsu Irrigation Trunk Canal, extending from Lake Hungtse at Kaoliangchien 105 miles northeast to the sea at Pientankang, was completed in 1952. Another important project, a dam regulating the Chungshan River outlet of Lake Hungtse, was completed the following year. Several major reservoirs have been completed on upstream tributaries of the Huai, including those at Paisha, Shihmantan, Panchiao, Pushan, Tapoling, Hsianghungtien, Fotzeling, Nanwan, and Meishan.

In the northernmost part of the Huai River lowland, on the Kiangsu-Shantung border, new outlets were built in 1952 for the I River and the Shu River. Both of these rivers descend in virtually parallel courses from the Shantung upland and often have caused floods in northernmost Kiangsu for lack of direct outlet to the sea. Together they drain an area of 15,000 square kilometers (5,800 square miles). Since the establishment of communes, a number of reservoirs and ponds have been constructed. It was reported that during the high-water season in the summer of 1963, these water conservation projects took 44 percent of the flow of the I River and 30 percent of the flow of the Shu River, and that a serious flood was thus avoided despite the unprecedentedly heavy rainfall.[9]

A series of water conservation projects to regulate the three major tributaries of the Huai River—the Pi, Shih, and Hang rivers—began in 1959 in the Liuan Special District, Anhwei. The area lies in the region of hills and mountains that constitutes the watershed between the Yangtze and Huai rivers, and used to be constantly under the menace of floods and droughts. Several big reservoirs and more than 2,000 kilometers of canals have been built as a part of the project for harnessing the Huai and its tributaries. The Shuch'eng-Luchiang Canal, 80 kilometers long, is the major part of the Pi-Shih-Hang irrigation system. By the end of 1969, the Pi-Shih-Hang irrigation system, which is one of the biggest in China, had brought more than 5 million mou of farmland under irrigation.

The construction of a 127-kilometer canal on the northern Anhwei plain began late in 1966 and was completed in 1969. The canal's purpose was to reduce the menace of flood and waterlogging in the Peinho (Toho) River basin, which drains 22,800 square kilometers in Honan, Kiangsu, and Anhwei. With its source in Honan Province, the Pienho River, a tributary of the Huai River, passes through Anhwei and empties into Lake Hungtse in Kiangsu. In the past, the Pienho would reach flood level during the rainy season. The completion of the "New Pienho River," as the canal is commonly called, has enabled the water to converge from an area of more than 6,900 square kilometers and flow to Hungtse Lake through the man-made waterway.[10]

The Hai River

The Hai River itself is only a short stream flowing 30 miles from Tientsin to the sea, but it serves as the outlet for five major rivers that drain the north China plain and converge, fanlike, at Tientsin. These are the Northern Canal, a section of the Grand Canal formed at Tungchow, east of Peking; the Yungting River, rising in northern Shansi as the Sangchien River; the Tach'ing River, formed in central Hopei; the Tzuya River, formed in south-central Hopei by the junction of the Huto and Fuyang Rivers; and the Southern Canal, a section of the Grand Canal that receives the Wei River from northern

Honan. In high-water seasons the flow on the rapidly rising upper reaches of the tributaries is far more than the Haiho can cope with. Several disastrous floods of the Hai River occurred in pre-Communist days. In 1917 and 1939 the floods breached the dikes and inundated most of the metropolis of Tientsin and vast areas of farmland in Hopei.

The multi-purpose Kuant'ing reservoir was completed in 1953 on the Yungting River, the northernmost of the five tributaries. The reservoir is located at a narrow gorge below the confluence of the Yungting and Kueishui rivers, 47 miles northwest of Peking. It impounds 80.131 billion cubic feet (2.27 billion cubic meters) of water, creating a lake of 85 square miles (220 square kilometers). The reservoir has made significant accomplishments in regulating the once widely fluctuating flow of the Yungting, which was 176,000 cubic feet per second at its peak and 35 cubic feet per second at the minimum, and in reducing the silt content of the river, which used to rise to as much as 40 percent of the volume of the water. But the reservoir has not eliminated the danger of flood, as evidenced by the overflow of the Yungting's banks near Lukouch'iao (the Marco Polo Bridge) on August 6, 1956, which inundated more than 7 million acres of farmland.[11]

The reservoir itself was reported to be silting up at such a rate that it affects energy output and water control as well as the irrigation systems. It was feared that a complete silting-up might take place in 60 years unless reforestation plans were set in motion at once.[12]

Scores of other reservoirs in addition to the Kuant'ing reservoir have been constructed on the upper reaches of the Hai River tributaries to hold large amounts of water in the summer and store water for the dry spring months; new outlets to the sea have been built on the lower reaches to give swifter passage for the floodwaters of the river. But these measures apparently were not enough. In 1963 unusually heavy rains swelled the Haiho and four of its five main tributaries spilled over, doing extensive damage to farms. In response to a directive issued by Mao Tse-tung in November of that year, the relevant departments of the Central Committee of the CCP and the State Council, together with the Hopei Provincial Party Committee and government, mapped out a new plan for harnessing the Hai River based on the principle of "detaining water in the upper reaches, dredging river beds in the middle reaches, and improving drainage in the lower reaches."[13]

The new harnessing scheme divided the Hai River and its tributaries into two major systems, southern and northern. The southern system consisted of the Heilungkang River, the New Tzuya River, the New Fuyang River, the Huto River, and the Tach'ing River; the northern system included the Yungting River, the Northern Canal, the Chaopai River, and the New Canal. During 1964-70 the work of the southern system was carried out by draining the Heilungkang basin, freeing more than 16 million mou (2.7 million acres) of farmland from floods; by building the New Tzuya River and the New Fuyang River, which drain 10,000 cubic meters of floodwater per second; by strengthening the northern dike of the Huto River; and by reinforcing the banks and deepening and widening the channel of the Tach'ing River. In addition 218 tributaries of the main rivers were dug and dredged, and more than 73,000 canals and ditches were constructed. In all these projects a total of 1.5 billion cubic meters of earthwork was dug.[14]

Following the completion of the southern system of the Hai River-harnessing project late in 1970, the work on the northern system was jointly undertaken by Hopei Province and the municipalities of Peking and Tientsin. Together the rivers of the northern system drain an area of more than 85,000 square kilometers.

The Pearl River

The Pearl River (Chu Kiang) delta is built up by the alluvial sediments of the three main rivers in Kwangtung and Kwangsi: the Hsi Kiang (West River), Pei Kiang (North River), and Tung Kiang (East River). This delta of less than 10,000 square kilometers, covering 11 hsien and the outskirts of three cities, is important in producing rice, sugarcane, silk cocoons, and jute. The low-lying alluvial land along the coast has suffered from river flooding, typhoons, rainstorms, and the backwash of sea tides, as well as from periodic drought. A number of dams, dikes, and tidal detention projects have been completed since 1949 to protect farmland and cities from floods and tidal waves, which often threaten the area in the wake of typhoons. A network of power transmission lines and 2,500 electric pumping stations with a total capacity of 180,000 kilowatts was completed in 1964 for irrigation and drainage purposes, and the construction of additional 2,000 stations with a total capacity of 58,000 kilowatts was planned. By the end of 1963, the network had been extended to cover 23 hsien and cities, with Canton as its center. It was expected that with the completion of the project at the end of 1964, more than half a million hectares of farmland on the delta and its adjacent areas would have mechanized irrigation and drainage. But in the summers of 1959 and 1966, as a result of heavy rain in most parts of Kwangtung, farmland was inundated in many areas, especially along the East River basin.[15]

The Sungari and Liao Rivers

A low watershed of 800 feet separates the Manchurian plain into the Sungari lowland in the north and the Liao lowland in the south. The Sungari, known in Chinese as the Sunghua, flows northwest until it meets the Nun River (Nonni). It then turns sharply northeast and receives the Mutan River before joining the Amur on the Sino-Soviet border. In the southern part of the Manchurian plain, the Liao River flows east and south to the Gulf of Liaotung, west of the Liaotung Peninsula.

Because of the concentration of annual precipitation in July and August, the Sungari basin is subject to frequent summer floods. Heavy damage was reported from floods in the area in 1951, 1953, 1956, 1957, and 1960, despite the construction and repair of nearly 600 kilometers of dikes along the river. In 1956 a plan was mapped out for the construction of 11 multipurpose reservoirs and other flood-control projects along the Sungari and its tributaries.[16]

The Fengman reservoir, 15 miles southeast of Kirin, the capital of Kirin Province, was the only large reservoir in China before the Communists took over. It is located at the point where the Sungari River leaves the east Manchuria uplands in a rapids-strewn section. The Japanese built the multiple-purpose project during World War II. The dam is 3,700 feet long and 300 feet high. The hydroelectric plant had eight turbines of 70,000 kilowatts each, but by the end of the war only two turbines were in operation as a result of dismantling by the Soviet army in 1946. The reservoir sustained damage during the decisive battle between the Nationalists and the Communists in the northeast. During the First Five-Year Plan period, the Peking government reconstructed the power plant with a projected capacity of 567,000 kilowatts. In March 1954 a 220,000-volt transmission line was installed, linking the Fengman station with the industrial complex at Shenyang, a distance of 230 miles. The reconstruction project was completed in October 1959.[17]

More than 200 reservoirs and 1,000 kilometers of dikes have been built along the West Liao River, a tributary of the Liao River, to control its periodic floods. The river's silt-laden water has been used to transform one-third of a million hectares of sandy land along its banks.[18]

In addition to the projects for the principal rivers, flood-control measures also have been taken for local water bodies. Nevertheless, floods and waterlogging remain a major menace to agricultural production. According to Chang Tzu-lin, president of the Research Institute of Water Conservation and Hydroelectric Power:

> This is because the standards for the prevention of floods along the big rivers are not high enough, and large numbers of medium-sized and small rivers have yet to be brought under control. Along the southeast coast, agricultural production is still menaced by tidal waves. The potentials of the existing engineering projects for the

control of floods, the elimination of waterlogging and the resistance to tidal waves have yet to be carried out.[19]

IRRIGATION

A large proportion of the cultivated land in China is irrigated, about 15 percent in north China and nearly 70 percent in the rice-growing regions. In both the wheat and rice regions, there is urgent need for a larger and more dependable supply of water. The NPAD provided that paddy fields and irrigated land should be expanded from more than 390 million mou (65 million acres) in 1955 to approximately 900 million mou (150 million acres) in 1967. For the purpose of ensuring good harvests, the capacity of the irrigation facilities to resist drought generally should be raised to between 30 and 50 days, according to the varied conditions in the different localities. In places suitable for double-crop paddy fields, the capacity should be raised to between 50 and 70 days (Article 5).

As shown in Table 11.1, there was only a gradual buildup in irrigated acreage during 1949-57. However, a very large increase was reported from 1957 to 1958 and was attributed by Communist authorities to the Leap Forward campaign. According to Vice-Premier T'an Chen-lin, the irrigated area had increased by 550 million mou (91.7 million acres) in the first two years of the Leap Forward campaign. This increase resulted in a total irrigated area of 1.07 billion mou (178 million acres), of which 610 million mou (102 million acres) were able to withstand drought for 30, 50, or 70 days, as required by the NPAD.[20] In another account Vice-Minister of Agriculture Ho Chi-feng claimed an increase of total irrigated area in China from 230 million mou (38.5 million acres) in 1949 to more than 1 billion mou (167 million acres) in 1959, or from 16.3 percent to more than 60 percent of the total cultivated area. According to his account, this meant that China's irrigated area had reached more than one-third of the world's total irrigated area, making China one of the most developed countries in water conservation.[21]

But several later statements have discredited the claims about total irrigated areas for 1958 and 1959. For example, the communiqué of the Ninth Plenary Session of the Eighth Central Committee of the CCP, held on January 14-18, 1961, said that the area "effectively irrigated" had increased by more than 300 million mou in 1958, 1959, and 1960.[22] On that basis and on the assumption that 520 million mou of irrigated area in 1957 were all in the "effective" group, the total irrigated area in 1960 would have been only 820 million mou.

Also, at the end of 1963 a People's Daily editorial stated that irrigated land at that time was only one-third of the cultivated land,

TABLE 11.1

Irrigated Area in the PRC, 1949-66

Year	Total Area Under Irrigation (million mou)
1949	240
1950	250
1951	280
1952	320
1953	330
1954	350
1955	370
1956	480
1957	520
1958	1,000*
1959 (planned)	1,100*
1960	820
1961	n.a.
1962	550
1963	510
1964	500
1965	520
1966	560

*See explanation in the text.

n.a. = not available.

Sources: For 1949-58, State Statistical Bureau, Ten Great Years (English ed.; Peking, 1960), p. 130; for 1959 Vice-Premier Li Fu-ch'un, "Report on the Draft Plan for the National Economy, 1959," presented at the First Session of the Second National People's Congress, April 21, 1959, Hsin-hua Pan-yueh-k'an [New China Semi-monthly], Peking, No. 9, 1959, pp. 15-20; for 1960, communiqué of the Ninth Plenary Session of the Eighth Central Committee of the CCP, January 14-18, 1961, JMJP, January 21, 1961, p. 1; for 1962 and 1963, "Actively Do a Good Job in Water Conservation for This Winter and Next Spring," editorial, JMJP, November 30, 1963, p. 1; for 1964, Economic Reporter, Hong Kong, No. 895, November 16, 1964, p. 20; for 1965, JMJP, September 30, 1965, p. 2; for 1966, JMJP, September 30, 1966, p. 3.

which was equivalent to a little over 500 million mou. This meant that there had been no notable progress between 1957 and 1963, since most irrigation works during that period were directed at improvement of existing projects, repairs, and maintenance.[23]

The 1958 and 1959 claims also contradicted the statement made by Su Tsung-sung, director of the Research Institute of Irrigation of the Chinese Academy of Agricultural Science, that the total area effectively irrigated in 1964 was approximately 500 million mou.[24] The total irrigated area in 1965 was reported to have increased 20 million mou over that of 1964, and the increase in the total irrigated area in 1966 was reported to have doubled the increase in 1965.[25] Based on Su Tsung-sung's figure for 1964 (500 million mou), the irrigated area would have been 520 million mou in 1965 and 560 million mou in 1966. These figures compare with the 1 billion mou claimed for 1958 and the 1.1 billion mou planned for 1959.

At present, irrigation in the PRC still depends mainly on human and animal labor, while only a small portion is operated mechanically (by diesel, gas, or steam engines) or electrically. In September 1965 "power-driven irrigation and drainage systems" served a total area of 84 million mou, a little over 16 percent of the 520 million mou irrigated and drained by all means in that year.[26]

The 2,200-year-old Tuchiangyen irrigation system on the Chengtu plain in Szechuan Province has been renovated. The skillfully designed multipurpose engineering project was intended to divert the flows of the Min River, a tumultuous stream that rises on the high plateau of Tibet. By building a series of dams and dikes on the main river where it first enters the broad plain from the mountain canyon, the farmers divided its flow into many parts so they could irrigate 500,000 fertile acres. The structure—composed of bamboo frames weighted down by rocks—also greatly reduced the heavy toll of life and property taken by spring and summer floods. As a result of constant efforts made since 1949 to renovate and expand the system, the irrigated area has reached more than 466,000 hectares, compared with 133,000 hectares in the early days of the Communist regime; and 27 hsien and suburban areas now benefit, as against 12 previously.[27]

During 1960 and 1961 the government set up more than ten electric irrigation areas, each embracing more than 1 million mou (167,000 acres). The electric irrigation areas were mostly in regions producing grain and cotton where adequate electric and water sources were available, in vegetable-producing areas around the cities, or in the industrial and mining districts. These included (1) the rice-producing regions of the Pearl River delta in Kwangtung, southern and northern Kiangsu, and the Hangchow-Kashing-Huchow area in Chekiang; (2) the important grain and cotton regions around Tientsin, Tangshan, Shihchiachuang, Paoting, and Changchiak'ou in Hopei, the Kuanchung area in Shensi, the area in northern Honan, the areas around Chengchow, Loyang, and Sanmen Gorge in Honan, the areas along the Shih River, the Pi River, and the Hangpu River in Anhwei, and the area in southern and central Shansi; and (3) the vegetable-producing regions in the suburbs of Peking, Shanghai, and Tientsin.[28]

Since the mid-1960s the government has placed increasing emphasis on the use of electric pumping facilities for irrigation and drainage. In 1964 there were 4 million horsepower of mechanical pumping facilities in the rural areas, as against 2.04 million horsepower of electric pumping facilities. Thus the ratio was about 2:1 in favor of mechanical pumping facilities.[29] But since then electric pumping facilities have increased more rapidly. It would seem that in the 1970s or so electric irrigation and drainage will catch up with mechanical irrigation and drainage, mainly because of the lower cost of the electric methods.

In addition to modern mechanical and electric pumping facilities, many traditional types of irrigation and drainage equipment were improved or semimechanized during the 1950s and 1960s. An improved type of waterwheel has played an increasingly important role in irrigating and draining paddy fields. The conventional type of waterwheel used in south China consists of a chain of wooden paddles and a wooden trough, the lower end of which is put in water. As the chain rotates, paddles pass through the trough and raise the water. The new type of waterwheel replaces the wooden trough with a round or square tube and reduces the number of paddles. The paddles are of the same size as the inner diameter of the tube; and as they pass through, the tube below the plates becomes a vacuum, causing the water to rise as in a pump. The new waterwheel is said to be two or three times more efficient and to cost and weigh two-thirds less than the traditional type. It can be made in various sizes and materials (wood, bamboo, plastic, or steel) and in different forms for operation by hand, treadle, a draft animal, or a motor. It can be installed and operated almost vertically, whereas the old type can work only at a slight angle.[30]

The use of a new type of turbine pump has been popularized in many areas in the south and southwest for lifting water to the upland fields. In the Kwangsi-Chuang Autonomous Region and in Fukien, Szechuan, Hunan, and Kweichow provinces, the improved turbine pumps are said to have enabled some communes to keep their paddy fields watered during an unusually severe dry spell.[31] Composed of a water turbine and a pump, the turbine pump consumes neither electricity nor oil, and is capable of lifting water to higher levels through water pressure. In addition to irrigation, it is used in supplying drinking water, generating electricity, and processing agricultural products. Turbine pumps were first installed in China's hilly areas in 1956 and were popularized in 1964. In moutainous areas of Hunan Province alone, their number increased from 370 in 1963 to more than 8,300, capable of irrigating some 1.1 million mou of farmland, early in 1966. In south China's hilly provinces there were 43,000 turbine pumps in 1966, irrigating more than 3 million mou of farmland.[32]

Raising water for irrigation by turbine pumps is less expensive than by waterwheels, internal combustion engines, or electric motors; but the turbine pump can be used only where a rapid flow of water is available to produce sufficient power. Therefore, it cannot be used for lifting water at rest.

Other kinds of pumps currently in use include chain pumps, centrifugal pumps suited for irrigation in river-covered areas in the south, hydraulic pumps operated by water power, three-part serial pumps for raising water uphill, deep well pumps of various types, submersible electric pumps, internal combustion water pumps, and dredgers that also can be used for pumping water.[33]

Along the lower reaches of the Yellow River in Shantung Province, many siphons have been built since 1965 to draw silt for fertilizing some 200,000 mou of paddy rice fields as well as to draw water for irrigating the land. The project was first experimented at Lich'eng near Tsinan, with reportedly successful results, and later was extended to 16 other hsien along the Yellow River. Power-operated irrigation and drainage equipment has been installed near the siphons to draw water when siphons fail to obtain an adequate quantity during a dry season. The use of siphons for irrigation does not seem to have proceeded beyond the experimental state.[34]

For irrigation as well as for grain processing, new models of windmills have been developed and widely used in the coastal areas.[35]

OTHER MAJOR WATER CONSERVATION PROJECTS

In addition to the projects already discussed, many other important ones involving both irrigation and flood control have been planned or carried out. One of these was to draw southern waters to the north. The annual volume of water in China totals about 2,680 cubic kilometers (95 trillion cubic feet). It was estimated that approximately 500 cubic kilometers (17.65 trillion cubic feet) were needed for irrigation in 1958 and 700 to 800 cubic kilometers (24.71 to 28.24 trillion cubic feet) in 1960. Therefore, for China as a whole water resources are more than sufficient for irrigation uses. However, the situation varies among the different regions. The water in the basins of the Yellow River, the Huai River, and the Hai River is far from sufficient for irrigation uses. In the south, on the other hand, water is abundant, more than sufficient for developing cultivated and wasteland in the area.

Thus a natural solution is to draw water from the rivers in the south, particularly the Yangtze and its tributaries, to the rivers in the north, including the Yellow River, the Huai River, and the Hai River. Work on a preliminary plan to draw water from the upper reaches of the Yangtze to the Yellow River through a series of canals and channels has been started by the Yellow River Water Conservation Committee of the Ministry of Water Conservation. A similar plan to draw the water of the Sungari River to the Liao River has been studied jointly by two surveying and designing institutes in Harbin and Shenyang; in this case the direction is from north to south.[36]

Meantime, it has been realized that tremendous amounts of money, materials, and labor would be needed to direct southern waters to the north. Furthermore, although master plans of water conservation for some of the individual rivers concerned already have been completed, they would have to be revised according to the changing conditions, while plans for other rivers, including the Yangtze and the Sungari, must be completed before the entire program to draw southern waters to the north can be started. Therefore, the authorities have urged that, at least for the time being, as much water as possible from the individual rivers should be used for irrigation purposes in the respective basins.

Another plan frequently discussed has been to increase the use of groundwater for irrigation. Many areas in north China and northwest China are subject to frequent threat of drought because of little precipitation and lack of surface water. However, some of these areas are well-endowed with deposits of groundwater that, if properly utilized, could extend irrigation considerably.[37]

In considering this plan the Communist authorities may not have taken into account the possibility that the water might contain too much salt. However, it was admitted that in Hopei Province, as more wells were dug, less and less underground water was obtainable from each well. In any case, like the plan to draw water from the south to the north, the plan for increasing use of groundwater has reached only a preliminary stage because of financial and technical difficulties.

In the 1973 water conservation year (October 1, 1972-September 30, 1973) a mammoth well-drilling campaign was launched to combat a prolonged drought in north China. The unprecedented scope of well drilling was one of the outstanding features of the year's work. Altogether 300,000 new pump wells were reported to have been completed in 14 provinces, autonomous regions, and municipalities, the greatest number of pump wells sunk in China in a single year. Various kinds of drills were made by state- and commune-run factories, including light-weight drills that can be conveniently taken across mountains. The communes overcame many difficulties with state aid and succeeded in sinking wells as deep as 100 or 200 meters. Many wells were completed not only on the plains but also on loess highlands and in limestone areas where drillers had failed in previous attempts to strike water. Even in the driest period of the year, the pump wells provided an unfailing supply of water. Hopei Province reported its best grain harvest in history, while in neighboring Shansi Province grain output increased considerably over 1972.[38] The total number of pump wells in China rose from nearly 200,000 at the beginning of the

Cultural Revolution to 1.2 million at the end of 1973, irrigating about 6.6 million hectares of farmland.

One of the most publicized water conservation projects was the construction in 1958 and 1959 of a canal network on some 40 million mou (6.7 million acres) of cultivated land in the area north of the Huai River in Anhwei Province. The "canalization" project was carried out in view of the extraordinary floods in that area in 1950, 1952, and 1956 and the equally disastrous drought in 1957. The project consisted of the construction of 9 canals, 2,621 large ditches, 15,732 medium-size ditches, and 111,074 small ditches. The project was scheduled to be completed at the end of 1959. It was expected that "canalization" would improve the irrigation of some 42.7 million mou (7.1 million acres) of cultivated land in the area, prevent floods and droughts, and provide facilities for navigation and for raising fish and aquatic plants.[40]

Eight other provinces (Hopei, Honan, Shantung, Shansi, Liaoning, Kirin, Heilungkiang, and Shensi) and the Autonomous Region of Inner Mongolia also were reported to have mapped out plans for the construction of canals and ditches in a total area of 366 million mou (61 million acres), to be completed by the end of 1959.[41] No details on the plans and their implementation have been revealed. The "canalization" project apparently did not produce good results, as evidenced by the serious floods and droughts in these provinces in 1960. In fact, because of improper construction and lack of a drainage system, the project has resulted in salification and alkalization of land in many areas.

One of the outstanding water conservation projects completed in the 1960s was the Red Flag Canal in Lin hsien, Honan Province. Mountainous and cut up by gullies, and having thin and stony soil, Lin hsien used to be an area of acute water scarcity. Construction work on the project began in 1960 and was completed in 1969. The entire network consists of a trunk canal, three branch canals, and numerous channels and ditches—altogether 1,500 kilometers of waterways. The unruly Chang River, lying northwest of Lin hsien, was dammed to raise its water level and to direct it through the Red Flag Canal to serve the entire hsien. Hundreds of small reservoirs were constructed along the canal. The network is reported to be able to irrigate 600,000 mou (100,000 acres) of farmland, and the water famine is said to have become history.[42]

METEOROLOGICAL AND HYDROLOGICAL WORK

As an important part of the program to combat natural calamities, increasing attention has been given to the improvement of meteorological and hydrological work. The NPAD called for the completion of the networks of meteorological observatories and stations and hydrological survey stations in the 12 years beginning in 1956 (Article 22). In 1965 the number of observatories and stations in the PRC reached 2,470, exceeding by nearly 50 percent the number in 1957. Also in 1965 observatories had been set up in more than 90 percent of the administrative districts (chou and leagues), and meteorological or weather stations had been established in more than 80 percent of the hsien throughout the country. Meteorological posts also had been set up in many communes.[43]

In March 1953 a team for studying agricultural meteorology was jointly organized by the Research Institute of Geophysics of the Chinese Academy of Sciences and the former North China Research Institute of Agricultural Science. This team was expanded in 1957 to become the Research Institute of Agricultural Meteorology under the Chinese Academy of Agricultural Science. The institute, in collaboration with the Central Meteorological Bureau under the State Council, coordinates the work of the meteorological offices and stations in rural areas throughout the country. The institute has conducted studies on the effects of climate on the growth of principal crops, the relationship of cultivation practices to climatic conditions, measures for the reduction of forest fires, the improvement of weather forecasting, and the development of simplified, low-cost meteorological apparatus for use on farms. In 1956 a unified method of meteorological observations for agricultural purposes was adopted by the Central Meteorological Bureau as a guide to farmers.

In accordance with the "walking on two legs" policy, both modern and old methods of meteorology were adopted. Although the Western calendar is used in daily life and for all official purposes, farmers are urged to observe the 24 chieh ch'i. These are the 24 periods of approximately 15 days each that make up a year according to the Chinese calendar. Each period is supposed to be characterized by certain climatic conditions. Farmers also are urged to note the adages concerning knowledge of climate that have been learned from centuries of experience and passed on from generation to generation.[44]

Since 1958 "supplementary weather forecasts" have been developed in various provinces as part of the agricultural meteorological network. These are local weather forecasts prepared by the hsien weather stations on the basis of the wide-area weather forecasts prepared by the major stations, but with revisions made according to local meteorological data and the experience of local peasants.[45]

Late in 1974 the New China News Agency reported that about 10,000 weather posts had been established in the countryside by communes and production brigades, with the assistance of state meteorological departments. In addition there were many widely scattered weather-watching groups and rain-measuring stations. Most of the weather posts are closely linked with the scientific experiment stations (or groups) in the countryside, and members of the related scientific research groups take charge of meteorological work. Some weather posts are located in middle or primary schools in the rural areas, with the teachers and students doing meteorological work on a part-time basis; others are located on fishing boats.

The rural weather posts, which are a new kind of establishment, have improved their services by means of on-the-spot meetings, study classes, and weather information consultation. At present many weather posts have small observatories for periodic checks on temperature, moisture, wind velocity, and precipitation. They constantly provide communes and production brigades with weather information and, on the basis of local weather changes, issue data to supplement weather forecasts made by state meteorological observatories (stations), thereby effectively improving the quality of the latter's services. They also have carried out investigations and studies, and have applied indigenous experiences in predicting weather changes to their forecasting. For example, in order to test the reliability of the indigenous experience that "ants moving their nests foretell a rain," the workers of the weather post of Hsinhua Commune at Pingchang hsien, Szechuan Province, made approximately 800 inspections of ants moving their nests. On the basis of these inspections, they wrote a 70,000-word record and discovered the law of the relationship between the ants' activities and subsequent weather changes. Over the past ten years they have collected more than 1,100 such experiences from the masses and thus have improved the accuracy of their forecasting.[46]

At the end of 1957, there were some 8,000 hydrological survey stations throughout the country. Although this total was reported to be an increase of 22 times over that in 1949, nearly one-half of the country was still without hydrological services in 1957. The Hydrological Bureau under the Ministry of Water Conservation, in collaboration with local stations, conducts surveys on water level, flow capacity, silting, precipitation, and evaporation of the major sources of water. In addition to compiling the current data, the bureau has reorganized and analyzed the materials prepared before the Communist take-over. Also, by the end of 1957 the bureau had established over 2,700 hydrological information stations for reporting and forecasting the water conditions of more than 100 rivers of various sizes.

In more recent years emphasis has been given to the strengthening of hydrological research, the expansion of the networks of hydrological survey stations and hydrological information stations, especially those operated by local governments, and the study of underground water in view of the importance of its use in agriculture.[47]

In line with the large-scale development of water conservation construction, the state had established nearly 10,000 hydrological survey stations by the end of 1974 along rivers and lakes; on mountains, in plains, and basins; and at reservoir and hydroelectric power station sites. In addition many rainfall forecasting stations had been set up and operated by the communes and the production brigades.

During the high-water season the flood-prevention organizations at the central, provincial, municipal, and regional levels, as well as the valleys of the Yangtze, Yellow, and Huai rivers, can supply information about the rainfall and other hydrological phenomena. In 1974 floods occurred on the I and Shu rivers, two tributaries of the Huai River. The hydrological survey stations along the rivers supplied timely and accurate hydrological information, making contributions to the antiflood efforts of the masses in Kiangsu and Shantung provinces.

As a result of the improvement of the hydrological equipment and techniques, the hydrological survey level has been greatly raised. For example, many hydrological survey stations now have remote-control equipment for the survey of river water flow and the discharging of river silt. The use of this equipment has shortened the working time and lessened the manual labor compared with the past, when wooden boats were used in survey work.[48]

THE PROSPECTS

Despite some improvement in water conservation work in recent years, many areas in the PRC are still vulnerable to periodic floods and droughts. Lu Wu, deputy director of the Central Meteorological Bureau, revealed in 1966 that calamities from floods and droughts had not been avoided in any year since 1949. The seasons and regions in which serious floods and droughts occurred during 1949-65 were given as follows:[49]

	Floods	Droughts
1949 summer	Haiho basin	
1950 summer	Huai River basin	
1954 summer	Yangtze River basin	
1956 summer	Yangtze and Huai basins	
1957 summer	Northeast China	
1957 summer-autumn		Huai River basin
1958 summer		South of the Yangtze
1959 summer		Yangtze and Huai basins

1959 spring	South China	
1959 summer	North China	
1960 spring		North China
1960 summer		Central China
1961 spring		North China
1961 summer		Central China
1963 spring		South China
1963 summer	North China	
1965 summer		North China
1965 summer	Huai River basin	

Table 11.2 shows the acreage of farmland affected annually by floods and droughts in 1949-61 and the number of persons so affected in 1949-56. The acreage of farmland affected during 1957-61 increased very rapidly, despite the claimed ability of the communes to "conquer nature through their collective strength and self-reliance." Table 11.3 shows the annual grain losses caused by floods and droughts in 1949-56, ranging from 2.0 percent to 6.3 percent of potential production. On the basis of the acreage of farmland affected, grain crops probably suffered greater losses from floods and droughts in 1957-61 than in 1949-56.

But it was not until the end of 1960 that the seriousness of floods and droughts was revealed. According to the official New China News Agency, the worst natural calamities in 100 years hit China in 1960 and 900 million mou (150 million acres), or more than half of the 1.6 billion mou of cultivated land, were affected with varying intensity. Of this total, 300-400 million mou (50-67 million acres) suffered heavily, and "a part of the farmland" did not produce anything.

The greatest havoc was caused by droughts, which affected 600 million mou (100 million acres). With the exception of Tibet and Sinkiang, all provinces and autonomous regions suffered from droughts at different times and with varying intensity. Hopei, Honan, Shantung, Shansi, Shensi, Inner Mongolia, Kansu, Szechuan, Yunnan, Kweichou, Kwangtung, Kwangsi, and Fukien were affected by droughts in both the spring and summer, while parts of Kiangsu, Anhwei, Chekiang, Hupeh, Honan, and Kiangsi were affected in the summer. More than 60 percent of the cultivated area in Hopei, Honan, Shantung, and Shansi was affected by droughts lasting for seven or eight months. In some areas, such as Linhsien and Ch'angkeh in Honan, the droughts lasted for more than a year. During the worst period of the droughts, eight of the dozen principal rivers in Shantung and certain sections of the lower reaches of the Yellow River had virtually no water.

Twenty provinces and autonomous regions were struck by typhoons and floods in 1960, most seriously in Liaoning, Kirin, Heilungkiang, Kwangtung, Fukien, Kiangsu, and Shantung. Eastern Liaoning

TABLE 11.2

Farmland Area and Number of Persons Affected by Floods and Droughts, 1949-61

Year	Farmland Area Affected (thousand mou)			Number of Persons Affected (thousands)		
	Total	Floods	Droughts	Total	Floods	Droughts
1949	127,870	127,870	n.a.	45,550	45,550	n.a.
1950	76,830	70,650	6,180	33,840	32,050	1,790
1951	56,630	22,140	34,490	30,340	10,440	19,900
1952	66,490	27,660	38,930	27,600	10,710	16,890
1953	58,090	47,970	10,120	17,250	14,760	2,490
1954	177,640	172,750	4,890	63,150	62,230	920
1955	108,100	46,090	62,010	24,950	13,530	11,420
1956	230,000	170,000	60,000	70,000	n.a.	n.a.
1957	200,000	n.a.	n.a.	n.a.	n.a.	n.a.
1958	570,000	100,000	470,000	n.a.	n.a.	n.a.
1959	650,000	n.a.	n.a.	n.a.	n.a.	n.a.
1960	900,000	300,000	600,000	n.a.	n.a.	n.a.
1961	900,000	n.a.	n.a.	n.a.	n.a.	n.a.

Note: n.a. = not available.

Sources: For 1949-55, Hsü K'ai, "Irrigation Work in China," Chung-kuo Shui-li [China Water Conservation], No. 10, 1956, pp. 5-18 (there are considerable discrepancies between the figures for 1953-55 given here and those given by Chou Po-p'ing in "State Purchases and State Sales of Grains Should Not Be Opposed," Liang Shih [Grains], Peking, No. 7, 1957, pp. 1-4); for 1956, Chou En-lai, "Report to the Fourth Session of the First National People's Congress, June 26, 1957," Hsin-hua Pan-yueh-k'an [New China Semimonthly], No. 14, 1957, pp. 1-16; Wang Kuang-wei, "Simultaneous Development of Industry and Agriculture, with Heavy Industry as Foundation," in 1956-1967-nien Ch'uan-kuo Nung-yeh Fa-chan Kang-yao (Hsiu-cheng Ts'ao An) Chiang-hua [Lectures on the National Program for Agricultural Development, 1956-1967 (Revised Draft)] (Peking, 1958), p. 21; for 1957 Hsü Ti-hsin, Chung-kuo Kuo-tu Shih-ch'i Kuo-min Ching-chi ti Fen-hsi [An Analysis of China's National Economy During the Transitional Period] (rev.; Peking, 1959), p. 129; for 1958, Chu Hsien-li, "In Celebration of the National Agricultural Exhibition," Nung-yeh K'o-hsueh T'ung-hsun [Agricultural Science Bulletin], Peking, No. 2, 1959, pp. 44-45, 56; for 1959, JMJP, January 23, 1960, p. 2; for 1960, JMJP, December 29, 1960, p. 1; for 1961, Ta Kung Pao, Hong Kong, April 1, 1964.

suffered the worst floods in the hydrological record of the province. The floods were aggravated by 11 devastating typhoons, the number being the highest for a single year in the previous 50 years.[50] It was noticeable that the haphazard and rash handling of water conservation work had contributed to the deterioration of water and soil conditions and had intensified the droughts and floods. Many of the projects did not perform their functions as satisfactorily as expected because of improper planning, lack of coordination, and poor quality of construction.

TABLE 11.3

Grain Losses Caused by Floods and Droughts, 1949-56

Year	Grain Production (million chin)	Grain Losses (million chin)	Losses as a Percentage of Potential Production
1949	216,200	11,400	5.0
1950	249,400	5,200	2.0
1951	270,100	6,300	2.3
1952	308,800	n.a.	n.a.
1953	313,800	15,000	4.5
1954	320,900	17,700	5.2
1955	349,600	12,800	3.3
1956	365,000	24,400	6.3

Note: n.a. = not available.

Sources: State Statistical Bureau, Ten Great Years (English ed.; Peking, 1960), p. 119; Hsin-Hua Pan-yueh-k'an [New China Semimonthly], No. 2, 1958, p. 109; Chou Po-p'ing, "State Purchases and State Sales of Grains Should Not Be Opposed," Liang Shih [Grains], Peking, No. 7, 1957, pp. 1-4.

While making fabulous claims, Vice-Minister of Agriculture Ho Chi-feng admitted that although nearly two-thirds of the country's cultivated land had been equipped with irrigation facilities, irrigation could be carried out in a normal manner on less than one-third of this land. Many of the projects had not developed their usefulness in time or had begun to irrigate with only low resistance to drought.[51]

The situation apparently did not improve in more recent years; in 1962 the People's Daily revealed that only one-third of all cultivated land throughout the country was effectively irrigated and that the efficiency of and benefit from the water conservation projects had not been fully exploited because of improper planning, incomplete assembling of facilities, untimely maintenance and repair, and inadequate management.[52] In another editorial the paper revealed that although water conservation facilities were well managed on the Manass River in Sinkiang, west Urumchi, in the Lohui canal irrigation area in Shensi, and in the Ch'ienchin reservoir in Hunan, many other works were not well done and had very limited usefulness.[53]

Moreover, improper construction and mismanagement of water conservation projects have caused salification and alkalization of land in many areas. In the latter part of 1959, when 70 million peasants were reported to be working on the "canalization" project, some individuals issued warnings that this would bring disaster. But these people were called "rightists."[54] It was not until early 1961 that the damage was admitted. In an article in the People's Daily, Huang Ping-wei, director of the Research Institute of Geography, Chinese Academy of Sciences, stated: "A secondary cause of alkaline soil is to be attributed to improper action by man, chiefly to improper arrangement of irrigation."[55]

An article in the Kuang-ming Daily by Hsiung I and Wang Tsun-ch'i described the situation as follows:

> There are many regions in which one can find examples of the soil becoming alkaline because it was irrigated but was not drained at the time of drought or semi-drought. In some places, the water rose suddenly and irrigation was stopped. In other places, salt water lay on the land for a long time and improvement of the soil was not attempted; there are regions where arable land is continually shrinking and alkali soil spreading. Agricultural work is badly affected.[56]

A more specific statement on the degree of alkalization of soil was later given by the People's Daily, which said that owing to the wrong irrigation system, there was a considerable increase of alkaline land in northern Kiangsu, in the north China plain of Hopei, Shantung, and Shansi, in the plains along the Sungari and Liao rivers, in the valleys of Inner Mongolia, and in Shensi, Kansu, Chinghai, and Tibet.[57]

Since the failure of the "canalization," the overall policy for the improvement of water conservation programs has been to emphasize the organization and management of various irrigation and drainage projects, including the following aspects: designing, planning, installation, repair, maintenance, and financing. Special attention has been given to the completion of the constituent parts of certain existing projects to make them work more effectively and to link them into integrated systems. Also, a number of drainage projects have been carried out along with irrigation projects in order to reduce water-logging and to expedite desalinization, especially in the depression areas along the coasts of Liaoning, Hopei, Shantung, and northern Kiangsu and in the beach areas around Tungting, Poyang, Hungtze, and other principal lakes. In the meantime, because of the shortage of steel, cement, stone, timber, and other necessary materials, as well as financial and technical difficulties, the government has continued to depend heavily on small projects that can be built by the collective efforts of the communes. Mass campaigns have been launched throughout the country for sinking pump wells, dredging river courses, and draining floodwater. In the north wells of both traditional and semi-mechanical types remain the most important tools for irrigation, while turbine pumps are widely used in the south to bring water uphill. Since cement reinforcement and stone dikes are expensive, the old method of earth banking is adopted for controlling floods. Machines are seldom used for dredging rivers.[58] Very few large projects were carried out or planned during the late 1960s, as such projects would involve tremendous outlays of funds on the part of the central government.[59]

Despite the difficulties the PRC has had considerable success in recent years in both irrigation and flood control by depending mainly on local resources and by improving planning and management. Although "ordinary floods and droughts" had not been eliminated by the end of 1967, as scheduled in the NPAD, a sound foundation has been laid since then for the long-range water conservation work that is vital to agricultural production.

NOTES

1. Ho Chi-feng, "Glorious Achievements of Water Conservation in Rural China for the Past Ten Years," Shui-li Shui-tien Chien-she [Water Conservation and Hydroelectric Construction], Peking, No. 18, 1959, pp. 13-17.

2. Hunan People's Radio Station, December 14, 1971, 7:00 P.M.; Union Research Service, Hong Kong, Vol. LXVI, No. 6, January 21, 1972, p. 65.

3. "Harnessing China's Rivers," Peking Review, No. 49, 1974, pp. 15-19.

4. NCNA-English, Sanmen Gorge, Honan, April 13, 1957.

5. "Reconstructing the Big Sanmen Gorge Dam," Peking Review, No. 5, 1975, pp. 13-14.

6. "No Miracle is Impossible," China Reconstructs, No. 10, 1970, pp. 19-25.

7. "First Stage of Giant Water Control Project Completed," Peking Review, No. 11, 1974, pp. 10-11.

8. Yang Jen-chang, Ch'ang Chiang [The Yangtze River] (Peking, 1958), pp. 22-26.

9. NCNA, Tsinan, October 8, 1963.

10. NCNA-English, Hofei, Anhwei, May 15, 1966; NCNA, Peking, December 2, 1969.

11. Fang Tsung-tai and Jen Tao-heng, "A Study of the Overflow of the Yungting River and the Functioning of the Kuant'ing Reservoir," *Chung-kuo Shui-li* [China Water Conservation], Peking, No. 5, 1957, pp. 66-73.

12. Jean Messines, "Forest Rehabilitation and Soil Conservation in China," *Unasylva*, Rome, Vol. XII, No. 3 (1958), pp. 103-20.

13. NCNA-English, Peking, November 23, 1966.

14. "A New Chapter in the History of River Regulation—A Report on the Great Struggle by the People in Hopei Province to Regulate the Haiho," *Kuang-ming Jih-pao* [Kuang-ming Daily] Peking, November 18, 1970, p. 1.

15. *Yang-ch'eng Wan-pao* [Canton Evening Post], June 24, 1966, p. 1; June 26, 1966, p. 1; June 27, 1966, p. 1; NCNA-English, Canton, October 24, 1961; November 12, 1963.

16. Ch'iang Hsiao-ch'u (Secretary of the Heilungkiang Provincial Committee of the CCP), "Immediately and Fundamentally Control the Sungari," statement made at the Eighth National Congress of the CCP, September 1956, *Jen-min Shou-ts'e* [People's Handbook] (Peking, 1957), pp. 502-503; Chang Tzu-lin, "Universal Construction of Water Conservation Projects and Technical Transformation of Agriculture," *Shui-li yu Tien-li* [Water Conservation and Electric Power], Peking, No. 16, 1963.

17. *JMJP*, October 6, 1959, p. 1.

18. NCNA, Huhehot, June 15, 1963.

19. Chang Tzu-lin, op. cit.; Chang Tzu-lin, "Water Conservation and Agricultural Production," *JMJP*, August 29, 1963, p. 5.

20. T'an Chen-lin, "Strive for the Fulfillment, Ahead of Schedule, of the National Program for Agricultural Development," *JMJP*, April 7, 1960, p. 1.

21. Ho Chi-feng, op. cit.

22. JMJP, January 21, 1961, p. 1.

23. "Actively Do a Good Job in Water Conservation for This Winter and Next Spring," editorial, *JMJP*, November 30, 1963, p. 1.

24. *Economic Reporter*, Hong Kong, No. 895, November 16, 1964, p. 20.

25. *JMJP*, September 30, 1965, p. 2; September 30, 1966, p. 3.

26. Special NCNA report on the occasion of the sixteenth anniversary of the founding of the PRC, NCNA-English, Peking, September 16, 1965.

27. NCNA-English, Peking, September 28, 1972.

28. *JMJP*, December 23, 1961, p. 1.

29. NCNA, Peking, September 20, 1964.

30. Peking Review, No. 34, 1965, p. 23.

31. Ibid., No. 20, 1964, p. 29; JMJP, April 19, 1964, pp. 1, 4.

32. NCNA, Peking, June 11, 1966; NCNA-English, Changsha, June 2, 1966.

33. Liu Pang, "Paving a Way to Agricultural Mechanization in China," JMJP, April 13, 1966, p. 2.

34. JMJP, May 23, 1965, p. 1.

35. Wu Sheng-hsun, "Windmills Currently Used in China's Rural Areas," JMJP, June 22, 1964, p. 5.

36. "Certain Questions Concerning the Future Development of China's Water Conservation Work," a statement by the leader of the Soviet water conservation team at the National Conference on Hydro-electrics held at Tientsin in January 1959, Shui-li Shui-tien Chien-she, No. 6, 1959, pp. 36-42.

37. Ho Ch'ang-kung, "Investigate Ground Water Resources; Aid Agricultural Production," Hung Ch'i [Red Flag], Peking, Nos. 9/10, 1961, pp. 42-47.

38. NCNA-English, Peking, December 31, 1973.

39. Ibid., January 7, 1974.

40. Chang Tso-yin, "Canalization of Huaipei," Chung-kuo Nung-pao [Chinese Agricultural Bulletin], Peking, No. 15, 1958, pp. 3-6.

41. "Canalization—A Measure for Water Conservation on the Plains," editorial, Chung-kuo Nung-pao, No. 15, 1958, p. 2.

42. Lin Min, Red Flag Canal (Peking: Foreign Languages Press, 1974), pp. 1-3.

43. JMJP, April 23, 1965, p. 2.

44. Research Institute of Agricultural Meteorology, "Accomplishments in Agricultural Meteorology Since the Founding of the People's Republic of China," Nung-yeh Hsueh-pao [Acta Agriculturae Sinica], Peking, Vol. X, No. 5 (1959), pp. 353-59.

45. Tan Wen-kuang, "On the Development of Supplementary Forecasting by Minor Meteorological Stations," K'o-hsueh T'ung-pao [Scientia], Peking, No. 17, 1958, pp. 529-31.

46. NCNA-Chinese, Peking, October 25, 1974; Chu K'o-chen and Wan Min-wei, Wu-hou-hsüeh [Phenology], (Peking: Science Press, 1973).

47. Wang Tsu-p'ing, "Functions of Hydrology," 1956-1967-nien Ch'uan-kuo Nung-yeh Fa-chan Kang-yao (Hsiu-cheng Ts'ao-an) Chiang-hua [Lectures on the National Program for Agricultural Development, 1956-1967 (Revised Draft)] (Peking, 1958), pp. 155-61.

48. NCNA Domestic Service in Chinese, Peking, March 18, 1975.

49. Lu Wu, "China's Climate and Calamities of Floods and Droughts," JMJP, January 26, 1966, p. 5.

50. JMJP, December 29, 1960, p. 1.

51. Ho Chi-feng, op. cit.

52. "Actively Do a Good Job in Water Conservation for This Winter and Next Spring," editorial, JMJP, November 30, 1963, p. 1; "The Major Keys to Increasing the Efficiency of and the Benefits from Water Conservation Projects," editorial, JMJP, April 10, 1962, p. 1.

53. "Water Conservation and Stable and High-Yielding Farms," editorial, JMJP, August 5, 1964, p. 1.

54. JMJP, September 28, 1959, p. 9; December 30, 1959, p. 1.

55. Huang Ping-wei, "Several Problems to be Considered in Expanding the Acreage of Cultivated Land," JMJP, May 5, 1961, p. 7.

56. Kuang-ming Jih-pao, April 20, 1961, p. 2.

57. JMJP, November 8, 1961, p. 2.

58. Ibid., October 16, 1965, p. 2; October 18, 1965, p. 3.

59. "Rely on 500 Million Peasants for Management of Water Conservation," editorial, JMJP, March 11, 1966, p. 2; Liu Shen, "The Policy of Emphasizing Small Projects Should Be Continued in Managing Water Conservation," Ta Kung Pao, Peking, March 11, 1966, p. 3.

CHAPTER 12

FERTILIZATION

"Water decides whether there will be a harvest, and fertilizer decides whether the harvest will be good or bad," says an old Chinese farm adage. For centuries China has struggled to maintain a moderate level of soil fertility by the use of human and animal wastes. However, years of cropping and heavy erosion have greatly reduced the soil fertility. To rebuild depleted soils large quantities of chemical fertilizers are required in addition to the usual organic fertilizers. Under the "walking on two legs" policy, the Communist government has placed equal emphasis on the use of chemical fertilizers and organic fertilizers, but the latter group—including human and animal excrements, green manure, pond and canal sediment, ashes, and industrial wastes—remains the principal source of soil fertilizer.

In Honan, for example, chemical fertilizers are now extensively used on the 7.38 million hectares of cultivated land, but each hectare of farmland in that province still takes an average of 75,000 kilograms of farmyard manure per year. The commune members have various ways of obtaining the fertilizers they need: composting garbage and earth, collecting night soil and pig and cow dung, fermenting plant stalks under high temperature, and planting milk vetch and sesbania as green manure. Total acreage under green manure crops in Honan Province is about 530,000 hectares every year. In addition some communes and production brigades use bacteria fertilizers.[1]

The NPAD urged the agricultural cooperatives (now the communes) to produce locally as much as possible of the farm manure and chemical fertilizers they needed (Article 6). During the Leap Forward campaign almost every one in the countryside took a hand in helping to accumulate animal manure or manufacture chemical fertilizers in local plants. Factory and office workers, students, housewives, and members of army units from towns and cities joined the rural population in a massive campaign of manure collection and fertilizer production. On January 28, 1959, an order was issued to the communes to the effect that in the allocation of manpower for the winter work program, priority should be given to manure collection and fertilizer production over work on irrigation and deep plowing. "Manure can be found everywhere and fertilizers can be made from almost anything," said the People's Daily, exhorting the PRC's millions to heed the directive.[2]

ORGANIC FERTILIZERS

Hog raising has been encouraged as an important source of manure as well as meat, bristles, and hides. The NPAD required each household in the countryside to raise an average of 1.5 to 2 hogs by 1962 and 2.5 to 3 hogs by 1967, except in the Moslem areas, where for religious reasons the communes and their members were urged to raise sheep (Article 6).

The number of hogs at the end of 1949 and each year from 1952 through 1974 are given in Table 12.1. According to Vice-Premier T'an Chen-lin, 270 million hogs were raised in 1959 and there were 180 million hogs in the sties at the end of that year. On the basis of 120 million farm households in the country, the average number of hogs raised by each household during the year was more than two, and that of hogs in the sties at the end of the year was 1.5. Thus the target set for 1962 was reached ahead of schedule.[3]

Despite this claim there has been an almost constant shortage of manure; and the communes and their members frequently have been urged to raise more hogs through good care of sows, multiplying of shoats, improvement of fodder supply, proper maintenance of hog pens, and other measures. The shortage of fodder in particular has been a major problem in raising hogs as well as other domestic animals. Efforts have been made to improve the feeding of livestock by means of analysis of the chemical composition of feed from different parts of the country and study of methods to expand the cultivation of high-quality feed.

Since the early 1950s the number of hogs in the PRC has fluctuated, generally in response to changes of government policy on hog raising. During the early years of the Communist regime, when hogs were mostly raised by individual farmers, the hog population grew steadily from 58 million in 1949 to 102 million in 1954. It declined to 88 million in 1955 and again to 84 million in 1956, because of agricultural cooperativization. In 1957 the government restored the emphasis on private hog raising and the number of hogs rose sharply to 146 million. Communalization probably has resulted in a decline in the hog population, although official figures for both 1958 and 1959, like

TABLE 12.1

Number of Hogs at the End of the Year, 1949, 1952-59, 1961-75

Year	Number of Hogs
Highest before 1949	78,530,000
1949	57,750,000
1952	89,770,000
1953	96,131,000
1954	101,718,000
1955	87,920,000
1956	84,026,000
1957	145,900,000
1958	160,000,000
1959	180,000,000
1961-65	196,917,000
1966-67	210,000,000
1967	213,000,000
1968	213,000,000
1969	215,000,000
1970	220,000,000
1971	223,000,000
1972	231,079,000
1973	235,831,000
1974	239,193,000
1975	231,000,000

Sources: For the highest figure before 1949 and for 1949, 1952, and 1957, State Statistical Bureau, Ten Great Years (English ed.; Peking, 1960), p. 132; for 1953-56, Hsien-hua Pan yueh-k'an [New China Semimonthly], Peking, No. 17, 1956, p. 43; for 1958, Ts'ai Tzu-wei, "Accomplishments in Animal Husbandry in the Past Decade," Chung-kuo Hsu-mu-hsuch Tsa-chih [China Animal Husbandry Magazine], Peking, No. 10, 1959, pp. 289-93; for 1959, JMJP, January 23, 1960, p. 1; for 1961-74, Food and Agriculture Organization of the United Nations, Production Yearbook, 1973 (Rome, 1974), p. 212, and Production Yearbook, 1974 (Rome, 1975), p. 200 (all figures are FAO estimates); for 1975, NCNA-English, Peking, October 17, 1975.

many other figures for the Leap Forward period, showed substantial increases. Data for the years after 1959 are not available, but it has been admitted that the number of hogs in 1960 dropped disastrously to fewer than 40 million, partly because of collective raising. Following the crop failure of 1959-61, it was necessary to adopt the policy of simultaneous collective and private raising, with emphasis on the latter. This policy was continued until the end of the Cultural

Revolution, when collective raising again was stressed. However, it would appear that continued emphasis on private raising would be necessary in order to increase the hog population.[4]

Since the adoption of the new economic policy for the countryside in early 1971, collective hog raising has been greatly expanded and individual households have been encouraged to raise more hogs in many parts of the country. Collective hog farms run by communes or production brigades provide shoats to commune members; production teams share part of the green and coarse fodders produced and help them process it; households are compensated for the manure they turn over for collective use. All this is reported to have helped to promote hog raising.[5]

In addition to the accumulation of animal manure and night soil, the planting of green manure crops has been encouraged not only as a source of organic substances but also as a measure for water and soil conservation and to provide feed for animals. The communes have been encouraged to grow green manure on fallow land and land between rows of other crops, and to raise hydrophytic green manure in bodies of water. Such green manure crops as sweet clover, cow vetch, and milk vetch traditionally are cultivated on winter fallow land in the rice-growing Yangtze River basin. These crops have proved effective in improving the acid red soils in south China and the saline land in north China. In recent years they have been introduced in some areas in Kwangtung, Kwangsi, Shensi, and a few other provinces. In 1964 green manure crops were grown on more than 80 million mou of land, an increase of 1.37 times as compared with the acreage in 1952. The communes have been asked to further expand their production of these crops.[6]

In the past few years many communes of the Yangtze valley and other rice-producing areas in the south have popularized planting red duckweed as a source of manure and fodder from rice fields and other water surfaces. In 1973 the acreage under duckweeds in various areas in the south reached 6.5 million mou (nearly 1.1 million acres). It rose to nearly 10 million mou (1.7 million acres) in 1974 and showed another increase of almost 100 percent in 1975. Red duckweed, also known as green duckweed or purple duckweed, is a plant of the aquatic bracken family. It breeds fast and is widely adaptable and high in output. It can be used as manure as well as fodder and can be grown with less labor and at less cost.[7]

But the production of green manure crops has been limited by seed shortage, cultivation systems, and natural conditions. Moreover, more than one-half the acreage of land growing these crops is in Kiangsi, Hunan, Chekiang, and a few other provinces, and its distribution is very uneven.[8]

In the past few years some areas have begun making a new organic fertilizer known as ammonium humate, which is a compound organic fertilizer made of humic coal, peat, or lignite (containing humic acid and other organic substances) mixed with ammonium, potassium, sodium, or other compounds. Long-lasting and efficient, it helps increase yields and at the same time improves the soil and regulates its acidity and alkalinity, which no inorganic fertilizers can do. China abounds in humic coal and other resources needed to produce humate fertilizer, which is easily made and has been widely used on large tracts of land in many places. According to incomplete statistics more than 7 million tons were produced in the first half of 1975.[9]

Other materials suggested as sources of organic fertilizers include wastes from food processing and other industries.[10]

CHEMICAL FERTILIZERS

The highest annual production of chemical fertilizers in China before the Communist take-over in 1949 was 227,000 tons. In the period 1928-36 the average annual importation of chemical fertilizers into the country, excluding Manchuria and Taiwan, was about 25,000 tons.[11] Thus the use of chemical fertilizers in China before World War II was insignificant when compared with the vast agricultural area of the country.

Estimates of China's annual requirements for chemical fertilizers have differed widely, ranging from a few hundred thousand tons to as much as 35 million tons. The NPAD urged the central and local authorities to produce approximately 5 to 7 million tons of chemcial fertilizers annually by 1962 and 15 million tons annually by 1967. According to the "Report on Actual Implementation of the 1959 National Economic Plan," released by the State Statistical Bureau on January 22, 1960, chemical fertilizer was one of a few key industries whose output was far short of the goal for the final year of the Second Five-Year Plan (1962). The production of most other industries—including steel ingot, coal, electricity, metallurgical equipment, generating equipment, and metal-cutting machinery—was reported to have already either reached or exceeded the 1962 target. The 1959 output of 1,333,000 tons of chemical fertilizers, on the other hand, represented only 19-27 percent of the goal for 1962, which was between 5 and 7 million tons, or 9 percent of the target of 15 million tons set for 1967 in the NPAD. In view of the slow progress in 1958-59, the production target for chemical fertilizers in 1962 was revised from 5-7 million tons to 3-3.2 million tons.[12]

No actual figures for annual chemical fertilizer production for some years after 1959 have been released; and the figures for recent

TABLE 12.2

Chemical Fertilizer Production in the PRC, 1949-75

Year	Chemical Fertilizer Produced (tons)
Highest annual production before 1949	227,000
1949	27,000
1950	70,000
1951	129,000
1952	181,000
1953	226,000
1954	298,000
1955	332,000
1956	523,000
1957	631,000
1958	811,000
1959	1,333,000
1960	1,676,000
1961	1,431,000
1962	2,147,000
1963	3,000,000
1964	3,600,000
1965	6,400,000
1966	8,100,000
1967	n.a.
1968	n.a.
1969	n.a.
1970	14,000,000
1971	16,828,000
1972	19,874,000
1973	25,000,000
1974	n.a.
1975	30,000,000-35,000,000 (planned)

Note: n.a. = not available.

Sources: For the highest annual production before 1949 and for annual production in 1949-58, State Statistical Bureau, Ten Great Years (English ed.; Peking, 1960), pp. 97, 103 (all figures do not include ammonium nitrate); for 1959, JMJP, January 23, 1960, p. 1; for 1960, figure is derived by deducting the 1958 and 1959 figures from 3,819,000 tons, which was the total production for 1958, 1959, and 1960, Ta Kung Pao, Hong Kong, October 1, 1962, p. 15; for 1961, figure is based on the statement that "production in 1961 was 52 times more than that in 1949," ibid.; for 1962, figure is based on the statement that "the 1962 production of chemical fertilizers increased by 50 percent over that of 1961," NCNA, Peking, January 2, 1963; for 1963, "Keep on Winning the Victory," editorial, JMJP, January 1, 1964, p. 1; for 1964, "Premier Chou En-lai's Report on the Work of the Government, at the First Session of the Third National People's Congress," JMJP, December 31, 1964, pp. 1-2; for 1965, JMJP, December 18, 1965, p. 1; for 1966, JMJP, December 30, 1966, p. 3; for 1970, Edgar Snow, "Talks with Chou En-lai: The Open Door," New Republic, March 27, 1971, pp. 20-23; for 1971, "New Leap in China's National Economy," Peking Review, No. 2, 1972, pp. 7-9; for 1972, Economic Reporter, English Supplement No. 1, Hong Kong, 1973, pp. 4-5, and NCNA, Peking, December 23, 1972; for 1973, based on the statement that "Output in January-November 1973 was 25.8 percent higher than that in the corresponding period in 1972, the biggest increase in any year," NCNA, Peking, December 20, 1973; for 1975, Snow, op. cit.

years shown in Table 12.2 are based on the percentage increases or decreases over the preceding years, according to various Communist sources. After a slight increase in 1960, there was a decline in 1961 because of the withdrawal of Soviet assistance and the general industrial recession. Production in 1962 and 1963 was reported to have increased considerably as a result of the first-stage installation of the Wu Ching Chemical Plant, the second-stage installation of the Kirin Chemical Fertilizer Plant, and the trial production of the Canton Nitrogen Fertilizer Plant. Production continued to increase until the Cultural Revolution, during which very little information was released. It was not until 1970 that the target of chemical fertilizer production set for 1967 in the NPAD was reported to have been nearly fulfilled. According to Premier Chou En-lai in 1970 production was approximately 14 million tons, as compared with the 1967 target of 15 million tons.[13]

The fundamental question as to what kinds of chemical fertilizers should be produced in China has been discussed frequently. Field tests conducted by the National Agricultural Research Bureau before World War II indicated that the use of nitrogen fertilizers would give significant increase in yields in most areas. Phosphorus fertilizers were needed in fewer cases, and potassium fertilizers still less frequently.[14]

This question was again discussed at a national conference on fertilizer research held on August 1-9, 1957, under the auspices of the Chinese Academy of Agricultural Science. The participants agreed that practically all the soil in China is deficient in nitrogen; that red and yellow soils along and south of the Yangtze River and slightly acidic soil in northeast China are generally dificient in phosphorus; and that potassium fertilizers are effective for increasing the production of certain crops in some regions. The conference recommended that priority be given to the production of nitrogen, phosphorus, and potassium fertilizers, in that order.[15] Li Ch'ing-k'uei, president of the Chinese Society of Soil Science, estimated that 80 percent of the soil in China is deficient in nitrogen, 50 percent deficient in phosphorus, and 15 percent deficient in potassium.[16] Chang K'ai-yen, a member of an investigation team on phosphorus fertilizer in Szechuan and Hupeh sponsored by the Ministry of Chemical Industry and the Ministry of Agriculture in 1961, was of the opinion that the manufacture of both nitrogen and phosphorus fertilizers should be promoted.[17]

In 1957 the Chinese Academy of Agricultural Science set up a national network for experimentation with chemical fertilizers. The network was composed of agricultural research units in 27 provinces, autonomous regions, and municipalities. The major function of the network was to study the effective use of chemical fertilizers for different crops in different regions and on different types of soil. Experiments have been conducted on 19 crops. The results of the experiments were thoroughly discussed at a conference convened by the academy in March 1963.[18]

Prior to the Communist regime ammonium sulfate was the only chemical fertilizer produced in China, with the highest annual output estimated at 227,000 tons. Ammonium nitrate and calcium superphosphate were produced for the first time during the First Five-Year Plan, and ammonium bicarbonate, fused phosphate, ammonia water, and ammonium chloride during the Second Five-Year Plan. In 1962 nitrogen fertilizers produced in the PRC included ammonium nitrate, ammonium sulfate, ammonium chloride, calcium cyanamide, urea, ammonia water, and ammonium bicarbonate; phosphorus fertilizers included defluorinated phosphate fertilizer, superphosphate, and mixed nitrogen and phosphorus fertilizers; and potassium fertilizers included potassium sulfate, potassium chloride, and mixed nitrogen and potassium fertilizers.[19] Since 1962 successful trial production of triple superphosphate, calcium metaphosphate, granulated calcium superphosphate, and ammonium molybdate has been reported.

At present the PRC is concentrating on the production of nitrogen fertilizers, although attempts also are being made to develop phosphorus and potassium fertilizers. In the production of nitrogen fertilizers, the proportion of ammonium sulfate is diminishing while that of ammonium nitrate is increasing. This is because ammonium sulfate contains a sulfate radical that tends to harden the soil and because its production requires large quantities of sulfuric acid, which is much needed by other industries. In recent years production of liquid ammonia by small plants has been emphasized in order to reduce the cost of manufacturing and river transport.[20] But in 1970 ammonium sulfate remained the principal fertilizer produced in the PRC.[21]

The PRC also is paying considerable attention to the development of urea production. Urea is not only a highly effective nitrogen fertilizer but also an important raw material for manufacturing plastics, synthetic fibers, and pharmaceutical products, and a culture medium for the strain used in making high-grade feedstuffs. Furthermore, plants for urea production require relatively little stainless steel, of which there is an acute shortage in China. A new technique in manufacturing granular urea, developed by the Nanking Chemical Fertilizer Plant in 1969, is reported to have simplified the manufacturing and equipment, and reduced capital investment.[22]

Since ammonium chloride and pure soda can be produced economically by a combined process, and since the production of ammonium bicarbonate requires relatively little capital outlay or material in short supply, the PRC is also developing the production of these nitrogen fertilizers.[23]

In the pre-Communist days China's chemical fertilizer industry consisted mainly of two nitrogen fertilizer plants, one set up by the Japanese in Dairen and the other by the Yung Li Company, a Chinese private enterprise, near Nanking. Both produced only ammonium sulfate. During the First and Second Five-Year Plans (1953-62), these two plants were expanded and three other large nitrogen fertilizer plants were constructed in Kirin, Taiyuan, and Lanchou. By the end of 1969, more than 20 large chemical fertilizer plants had been built. More large and medium-size chemical fertilizer plants have come into being since the Cultural Revolution, although very little information about them is available. As of the latter part of 1975, most provinces, municipalities directly under the central authorities, and autonomous regions had their own large and medium-size chemical fertilizer plants.[24]

In addition to these large plants, many "native-method fertilizer plants" were built in the communes during the Leap Forward campaign. These native-method chemical fertilizer plants varied widely in size, ranging from little shacks with two or three workers and a few vats to plants capable of producing as much as 8,000 tons of liquid ammonia per year. The total production of the native-method plants is not known, but it may have constituted an important portion of the total annual chemical fertilizer production. Most of the fertilizers produced by these plants were of low grade, only one-tenth to one-twentieth as effective as those produced in modern chemical fertilizer plants, according to a survey made in 1958 in Hopei, Kiangsu, Anhwei, and several other provinces. In general the method of preparing chemical fertilizers in small local plants was to transform such materials as nitrate rocks, bones, seaweed, and ashes from grasses and trees into powder and liquids, and then to mix, boil, and drain these materials so that finally a fertilizer containing a certain amount of nitrogen, phosphorus, and potassium was produced. Generally the percentages of these elements were low. Sometimes, because of improper proportions of the materials used, the fertilizer was either too acid or too alkaline for the local soil. A similar problem existed in the preparation of compost fertilizers, which are made from various materials, including vegetable refuse, human and animal manure, ashes, bones, feathers, furs, industrial wastes, and mud taken from the bottom of lakes and rivers.[25]

Because of the low efficiency of the small native-method plants, an effort has been made since 1958 to transform as many of them as possible into small modern-method plants. A small synthetic ammonia plant with an annual output of 800 tons was constructed in 1959 by the Dairen Chemical Company as an example of the modernization of native-method chemical fertilizer plants. This plant was built under the guidance of a work team composed of representatives of the State Economic Commission, the State Construction Commission, the State Planning Commission, the Ministry of Agriculture, and the Ministry of Chemical Industry. The annual production of 800 tons of synthetic ammonia was believed to be sufficient to meet the nitrogen needs of some 200,000 mou (33,334 acres) of land. It was estimated that this type of plant could be built by a commune in from four to six months at a cost of 819,000 yuan (U.S. $335,000) and using 198 tons of steel. However, the cost of production of nitrogen was found to be high, 400 yuan (U.S. $163) per ton, as compared with 180 to 200 yuan (U.S. $74-82) per ton at the larger plants.[26]

The attempt to improve small nitrogen fertilizer plants remained unsuccessful until 1962, when a new process for producing ammonium bicarbonate containing 17.5 percent nitrogen was developed by Hou Te-pang (1890-1974), an American-trained engineer who also directed the operation of the Yung Li Ammonia Plant constructed near Nanking in 1937. The new process was first tested successfully at a chemical plant at Tanyang, Kiangsu Province; and in 1963 the plant produced 2,600 tons of synthetic ammonia that was used for making 10,000 tons of ammonium bicarbonate, exceeding the projected capacity by 30 percent.

The small plant is said to be superior in that it brings quicker results with less investment. It uses local resources to the full, rather than using coke and quality anthracite, as in conventional plants. The small plant also economizes on equipment, raw material, and electric power. Since most of the plants are located in the countryside, the product is within easy reach of the users. Because of their small size such plants can be built in large numbers by localities and there are no delays in supplying chemical fertilizers to the surrounding farms. Moreover, the small plant is easier to operate: most of the workers in the Tanyang Fertilizer Plant had received only a primary or junior middle school education.[27]

Because of the success at Tanyang, small nitrogen fertilizer plants using the new process have been constructed in Kiangsu, Anhwei,

Honan, Shantung, Shensi, Shanghai, and other important grain-producing provinces and municipalities. Their total number exceeded 100 by June 1966, and increased to 800 in 1972 and to more than 1,000 in 1973. The total production capacity of small nitrogen fertilizer plants increased from 2 percent of that of chemical fertilizer plants of all sizes in 1961 to 12.4 percent in 1965, 18 percent in 1966, 33 percent in 1968, 43 percent in 1969, more than 40 percent in 1970, 43 percent in 1971, 53 percent in 1972, 54 percent in 1973, and 60 percent in 1974.[28]

Between 300 million and 400 million mou of the cultivated land south of the Yangtze River are deficient in phosphorus, according to a survey made in 1964. To improve soil fertility, production and use of phosphorus fertilizers have been promoted in recent years in Kwangtung, Kwangsi, Yunnan, Chinghai, Chekiang, Szechuan, and Hunan. In 1964 the PRC had nearly 100 phosphorus fertilizer plants with an aggregate annual production capacity of 2 million tons.[29]

Kwangtung and Kwangsi have made the most significant progress in the production and use of phosphorus fertilizers. In 1963 phosphorus fertilizers accounted for 85 percent of all the chemical fertilizers produced in Kwangtung, and they were used on 80 percent of the cultivated land in that province. In the same year 25 of the 27 chemical fertilizer plants in Kwangtung produced mainly phosphorus fertilizers. The Chanchiang Chemical Industry Plant, with an annual production capacity of 100,000 tons of phosphorus fertilizers, was built in 1962 with state financing. Construction of the other 24 plants, with annual production capacities ranging from 3,000 to 15,000 tons, was for the most part financed by district or hsien governments.[30]

In the Kwangsi-Chuang Autonomous Region, the Luchai Chemical Plant started operation in 1965 with an annual production capacity of 200,000 tons of calcium magnesium phosphate. Other phosphorus fertilizer plants in the region had a total production capacity of more than 250,000 tons annually.[31]

In Yunnan the production and use of phosphorus fertilizers has been popularized since 1965. Several plants have been constructed, but the quality of their products is not high. Phosphorus fertilizers are used in most areas of the province in conjunction with farm manure. In 1964 the province produced more than 70,000 tons of phosphorus fertilizers, and their use was extended to 40 hsien.[32]

There are several serious problems in the production of chemical fertilizers in the PRC. First is the problem of inadequate equipment. The Kirin, Lanchou, and other fertilizer plants built during the First Five-Year Plan period were basically installed with imported equipment. The equipment for the two nitrogen fertilizer plants in Lanchou and Taiyuan was supplied primarily by the Soviet Union, while that for the Szechuan Nitrogen Fertilizer Plant was imported from

Czechoslovakia.[33] In more recent years the PRC has imported fertilizer plants mainly from Western European countries, including, in 1964, five synthetic ammonia plants from Great Britain, each with an annual production capacity of 160,000 tons; a urea plant from the Netherlands with an annual production capacity of 175,000 tons; and two urea plants from Italy with a total annual production capacity of 300,000 tons.[34]

Plans for the construction of heavy equipment for producing chemical fertilizers were mapped out at the end of 1957 as a part of the Second Five-Year Plan. The First Ministry of Machine Industry, the Second Ministry of Machine Industry, the Ministry of Electric Appliances, and the Ministry of Chemical Industry formed a committee including one vice-minister from each ministry, and technical personnel from the ministries were pooled for drawing up plans.[35] But details of the plans and their implementation have not been revealed.

Since 1959 increasing quantities of equipment for producing nitrogen fertilizers have been manufactured in China. Late in 1962 the PRC claimed to be able to independently design and construct fertilizer plants of different sizes and build large high-pressure compressors of 2,400-5,000 horsepower, high-pressure seamless steel pipes, and high- and super-pressure cast-steel containers. This development of plant equipment was said to have helped not only the chemical fertilizer industry but also the high-pressure organic synthetic industry.[36]

Early in 1963 more than 100 machinery and electrical equipment factories throughout the country were assigned to produce full sets of equipment for the nitrogen fertilizer industry These factories included boiler plants in Harbin and Shanghai, heavy machinery plants in Taiyuan and Canton, high-pressure valve factories in Shanghai and Shenyang, and air separating equipment factories in Kaifeng and Hangchou. However, most of these factories had to be remodeled before they could produce full sets of equipment for large nitrogen fertilizer plants (each capable of producing the 25,000 tons of synthetic ammonia per year required for an annual production of 100,000 tons of ammonium sulfate fertilizer).[37]

The first-stage installation of the Wu Ching Chemical Plant at Shanghai, completed in September 1963, was the first large nitrogen fertilizer plant designed and constructed by the PRC with Chinese-produced equipment. Technical personnel from more than 90 construction and installation units in Shanghai participated in the project; and more than 100 factories in the city cooperated to manufacture the equipment, which previously had been largely imported. The equipment installed, which reportedly was capable of producing 25,000 tons of synthetic ammonia annually, included a high-pressure compressor, a large ammonia converter, and many precision meters. The completion of the project was hailed as "a momentous victory of China in her guideline of self-reliance in construction." Several ministries of the central government jointly sent a committee to Shanghai to study the method of installation.[38]

Early in 1965 a new workshop for producing 40,000 tons of urea annually went into trial operation at the Wu Ching Chemical Plant. The workshop consisted of more than 200 machines, including a huge compressor and a 36-meter-high tower for recovering carbon dioxide. The largest machine weighed 120 tons. New types of stainless and alloy steel supplied by Chinese mills were used for resistance to high temperature, high pressure, and corrosion, according to China Reconstructs.[39]

With the completion of its third-stage construction in 1968, the Hopei Chemical Fertilizer Plant began large-scale production of urea. Designed, equipped, and installed by Chinese engineers, the plant was described as "having fully met advanced world standards with new processes, new catalysts and large modern equipment for producing urea and yet requiring only half as much investment, one-fourth as much plant area, and one-third as many workers as old-style factories."[40] The plant, which was reported to be one of the major projects in the Third Five-Year Plan, had begun producing ammonium nitrate earlier.[41]

Another important project using equipment designed, manufactured, and installed by Chinese technicians was the first large modern nitrogen fertilizer plant, built in Shensi Province, which was put into operation in 1969. The plant uses a carbonization process for producing ammonium hydrocarbonate, which is said to be simpler than other processes and to require less money and less time for construction.[42]

However, even with these reported accomplishments, the PRC is still facing a serious problem of providing adequate equipment for expanding production of chemical fertilizers.

Another serious problem is the shortage of funds for capital construction investment. In 1961, when the policy of "developing agriculture as the foundation of the national economy" was adopted, 41 percent of capital construction investment allowed to the Chinese chemical industry was allocated for the production of chemical fertilizer, a higher proportion than in any previous year.[43] Although the actual amount of funds allocated to chemical fertilizers was not revealed, it was obvious that this percentage was difficult to attain because in the 1960s the PRC also gave more attention to pesticides, plastics, textiles, rubber goods, enamelware, and other industries, many of which require the use of chemicals.

Another difficulty is that the available supply of steel, cement, timber, and other materials for the construction of fertilizer plants is far below the actual need. The shortage of raw materials and electric power for making chemical fertilizers is equally acute. The phosphorus mines at Chinp'ing, Kiangsu Province, were reported to have supplied 300,000 tons of pulverized rock phosphate annually, beginning in 1958.[44] Many other phosphorus mines of various sizes are reported to have been discovered during the 1960s, but large portions of the deposits have low phosphorus content and are not profitable for making phosphorus fertilizer.[45] In Chinghai the bones of sheep and cattle are used as the principal raw material for making phosphorus fertilizer.[46]

With these serious problems the PRC has been facing a very difficult task in expanding the production of chemical fertilizers to meet the increasing need. Table 12.3 shows that both supply and production of chemical fertilizers increased steadily from 1952 to 1958. But, because of the rapidly growing demand and only slow expansion of production, it has been necessary to import chemical fertilizers in increasing quantities. Although data for more recent years are not available, the deficit in chemical fertilizers apparently has remained acute, as evidenced by the large amounts of chemical fertilizers imported. In 1969 the PRC imported 6 million metric tons of chemical fertilizers at a cost of U.S. $180 million.[47] In 1970 it imported more than 8 million metric tons of chemical fertilizers, accounting for 20 percent of the world's total fertilizer imports, and became the world's largest fertilizer importer. Ammonium sulfate, the chemical fertilizer of which China produces the greatest quantities, has been the largest item among the imported fertilizers. The second largest item imported has been calcium ammonium nitrate. Imports of urea and ammonium chloride also have increased considerably.

In recent years Japan has become increasingly important as a supplier of chemical fertilizers to the PRC. In 1968 Japan's exports of chemical fertilizers to the PRC, in terms of ammonium sulfate, totaled 2,459,000 metric tons, with a total value of U.S. $74 million. (The weight of all chemical fertilizers is commonly calculated on the basis of their active strength as compared with ammonium sulfate.) In 1969 the total weight increased to 2,848,000 metric tons and the total value to U.S. $80.8 million, accounting for 20.7 percent of the total value of all Japanese exports to the PRC.[48] In 1970 Japan exported 5.47 million tons of chemical fertilizers to the PRC.[49]

Also important as a supplier of chemical fertilizers to the PRC has been Nitrex, an organization of ten leading nitrogen producers and exporters in Belgium, West Germany, France, the Netherlands, Italy, Norway, Austria, and Switzerland, with headquarters in Zurich. In the fiscal year 1967/68 Nitrex sold about 3 million metric tons of nitrogen fertilizers to the PRC with a total value of U.S. $100 million, while Japan supplied the PRC with 2.25 million metric tons at a total value of U.S. $70 million.

TABLE 12.3

Production and Supply of Chemical Fertilizers, 1949-58
(thousand tons)

Year	Production	Supply	Deficit
1949	27	n.a.	n.a.
1950	70	n.a.	n.a.
1951	129	n.a.	n.a.
1952	181	318	137
1953	226	592	366
1954	298	802	504
1955	332	1,255	923
1956	523	1,608	1,085
1957	631	1,944	1,313
1958	811	2,708	1,897

Note: n.a. = not available.

Source: State Statistical Bureau, Ten Great Years (English ed.; Peking, 1960), pp. 97, 171.

Because the price of Japanese fertilizers, including freight, is lower than that of European fertilizers, it is likely that the PRC will increase its purchases from Japan. The FOB price of European fertilizers in 1969 was $28 per ton, while that of Japanese fertilizers was about $30. But, as a result of the closure of the Suez Canal in 1968, the price of European fertilizers, including freight, has risen to considerably more than that of Japanese fertilizers. Moreover, Japan uses small ships to carry fertilizers to China and therefore can deliver the goods to any port. The date of delivery is punctually observed.[50]

Among other countries that have supplied the PRC with chemical fertilizers are Morocco, the Soviet Union, Great Britain, Finland, Kuwait, Jordan, Canada, and Chile.

Even if chemical fertilizers could be provided in sufficient quantities, there remains the question of how to use them efficiently. This requires technical knowledge that is not yet common in the communes. It has been suggested that specialized personnel be entrusted with the handling of chemical fertilizers, but the suggestion only showed that even the most fundamental forms of organization are still wanting.

There was some hesitation about the use of phosphorus fertilizer particularly. In 1956 China's first plant producing calcium superphosphate was completed in Nanking with Russian assistance, and during the Leap Forward campaign small phosphorus fertilizer plants were established in several provinces. However, the phosphorus fertilizer did not find buyers, especailly after the beginning of 1961. The low sale of phosphorus fertilizer was believed to have been caused by several factors. First, its use was ineffective because of the lack of a balance between phosphorus and nitrogen fertilizers. In Szechuan Province, for example, phosphorus fertilizer was distributed equally among the communes and was not concentrated in places where the soil needed it. Another important reason was that the fertilizer produced was not sufficiently good quality. In many cases it burned the seed and the growing plant, resulting in decreased production. There was no standardization and inspection of quality. In many communes the peasants regarded phosphorus fertilizer with great suspicion. They thought it was just pulverized stone. Also, the communes were reluctant to buy the fertilizer because its price was high.[51]

There was also difficulty in the use of liquid ammonia in certain areas of Shantung, Szechuan, Chekiang, and Kiangsu provinces, where sulfuric acid was in short supply. The farmers initially were repelled by the offensive smell of liquid ammonia, but they were patiently taught by special teams how to use it. Some peasants complained that once they handled the fertilizer, they would lose their sense of smell forever.[52]

It also was reported that in Liaoning, Shantung, and Hopei provinces, production teams spent much money buying magnesium products for fertilizing. A reader of the People's Daily called attention to this matter, explaining that Chinese soil needs very little magnesium and that it is very doubtful whether such fertilizer could be of any use.[53]

In Fangshan and Ch'angp'ing hsien of the municipality of Peking, some production teams used large quantities of ammonium bicarbonate to mix the seeds, with the result that the rate of germination was lowered. Other production teams spread the ammonium bicarbonate over the fields and exposed it to the air, so that it rapidly decomposed. Many peasants in the hsien complained that ammonium bicarbonate had killed the crops and had not shown any fertilizing effect.[54] Finally, the inadequacy of storage and transportation facilities has resulted in heavy damages and losses of chemical fertilizers in the PRC. The conditions of many of the warehouses are poor: the floors are damp, no protective measures are taken to separate the chemical fertilizer from the floors, and as a result much of the fertilizer is moistened and decomposed. The inefficient inland river transport system and loading and unloading methods also have caused considerable damage to chemical fertilizers. In some places in Kwangtung, the rate of damage and loss of chemcial fertilizers in the course of storage and transit was estimated at more than 7 percent in 1963.[55]

THE PROSPECTS

The production, distribution, and use of chemical fertilizers in the PRC seem to have improved somewhat in the past few years.

According to Chou En-lai, the PRC's chemical fertilizer production in 1970 was 14 million metric tons. This figure seems to be too high, even though it is still 1 million metric tons short of the goal for 1967 prescribed in the NPAD. He estimated the country's annual need for chemical fertilizers at 30-35 million metric tons, which was also the production goal for 1975, the final year of the Fourth Five-Year Plan.[56] For the estimated 1.65 billion mou of cultivated land, this is equivalent to 36-42 chin of chemical fertilizers per mou (273-320 kilograms per hectare), which would place the PRC with the Netherlands and Japan among the countries applying the largest amount of chemical fertilizers to the same acreage of cultivated land.

In recent years China has purchased a number of large fertilizer plants from the United States, the Netherlands, Japan, and France. Its imports of finished fertilizers are likely to decline in the future because of its increasing domestic capacity. But even if China's capacity for producing chemical fertilizers continues to improve, it would have been difficult to increase its annual production from the reported 14 million metric tons in 1970 to 30-35 million metric tons in 1975; that is, to increase production by an average of 3 or 4 million metric tons per year over five years.

Because of the periodic shortage of manpower and fodder for raising hogs and other difficulties, it is also doubtful that the supply of organic fertilizers, produced mainly by the communes, could increase sufficiently to meet the rising demand. Therefore, to make up the deficiencies, the PRC probably will have to continue to import chemical fertilizers in considerable quantities for many years.

NOTES

1. Hsiang Jung, "An Effective Way to Speed up Chemical Fertilizer Production," Peking Review, No. 32, 1975, pp. 16-19.

2. JMJP, January 29, 1959, p. 1.

3. T'an Chen-lin, "Strive for the Fulfillment, Ahead of Schedule, of the National Program for Agricultural Development," JMJP, April 9, 1960, p. 1.

4. JMJP, August 31, 1960, p. 1; Ta Kung Pao, Peking, August 31, 1961, p. 1; January 28, 1962, p. 1; June 12, 1962, p. 1.

5. NCNA-English, Peking, October 17, 1975; Shang Shih-wen, "Raise Hogs with Vigor and Extra Vigor," JMJP, November 20, 1975, p. 3; Livestock Bureau, Ministry of Agriculture and Forestry, Ta-li Fa-chan Yang-chu Shih-yeh [Strive to Develop Hog-Raising Enterprise] (Peking: Agriculture Press, 1974).

6. JMJP commentator, "Develop Cultivation of Green Manure to Insure Soil Fertility," JMJP, June 5, 1965, p. 2.

7. "Wrest Manure and Fodder from Paddy Fields and Water Surfaces," JMJP, July 17, 1975, p. 2.

8. JMJP, June 5, 1965, p. 2; Chang Hsin-i, "Grow More Green Manure Crops in South China," Hung Ch'i [Red Flag], Peking, No. 18, 1962, pp. 19-29.

9. Hsiang Jung, op. cit.

10. "Fertilizer Is the Key to Achievements of Continued Forward Leap in Agricultural Production," editorial, JMJP, July 19, 1960, p. 1; T'u Feng, "Fully Utilize Fertility Resources and Increasingly Accumulate Organic Fertilizers," Hung Ch'i, Nos. 9/10, 1961, pp. 53-57.

11. K. K. Yao, Fertilizers in China (Washington, D.C.: Food and Agriculture Organization of the United Nations, 1948), p. 19.

12. JMJP, January 23, 1960, p. 1.

13. Edgar Snow, "Talks with Chou En-lai: The Open Door," New Republic, March 27, 1971, pp. 20-23.

14. Office of Foreign Agriculture, U.S. Department of Agriculture, Report of the China-United States Agricultural Mission (Washington, D.C., 1947), pp. 5-6.

15. K'o-hsueh T'ung-pao [Scientia], Peking, No. 17, 1957, pp. 541-42.

16. JMJP, December 25, 1962, p. 5.

17. Chang K'ai-yen, "A Discussion of Certain Questions on the Promotion and Use of Phosphorus Fertilizers," JMJP, November 15, 1962, p. 5; Li Ch'ing-k'uei, "An Examination of the Problem of Phosphorus Fertilizers in China from the Viewpoint of Soil Fertility, Properties of Phosphorus Deposits and Agricultural Conditions," K'o-hsueh T'ung-pao, No. 24, 1959, pp. 820-23.

18. Kuang-ming Jih-pao [Kuang-ming Daily], Peking, March 22, 1963, p. 1.

19. "Chemical Fertilizer Industry Energetically Supports Agriculture," Ta Kung Pao, Hong Kong, October 1, 1962, p. 15.

20. Investigation Group of the Shanghai Municipal Revolutionary Committee, "Run Small Fertilizer Plants Well," Hung Ch'i, No. 12, 1969, pp. 36-42.

21. Snow, op. cit.

22. NCNA, Nanking, April 18, 1969.

23. Kao Kuang-chien, "Chemical Industry Serves Agriculture," Peking Review, No. 29, 1963, pp. 25-28.

24. Hsiang Jung, op. cit.

25. K'o-hsueh T'ung-pao, No. 21, 1958, pp. 656-57.

26. JMJP, July 19, 1960, p. 2; "Turn out More Chemical Fertilizers," editorial, JMJP, August 7, 1960, p. 1; Hua-hsueh Kung-yeh [Chemical Industry], No. 22, 1959, pp. 4-6.

27. JMJP, August 4, 1964, p. 1; "A Strong Reinforcement for Nitrogen Fertilizer Industry—Development and Growth of Small Nitrogen Plants amid the Struggle Between the Two Lines," JMJP, April 10, 1969, p. 4.

28. NCNA, Nanking, June 14, 1966; Ta Kung Pao, Peking, June 15, 1966, p. 1; "Growth of Small Nitrogen Fertilizer Plants," Peking Review, No. 24, 1969, pp. 33-35; NCNA-English, Peking, June 2, 1970; NCNA-English, Peking, January 1, 1970; "Fertilizer Industry Comes to the Countryside," China Reconstructs, Peking, Vol. XIX, No. 1 (1970), pp. 20-25; NCNA-English, Peking, December 27, 1970; NCNA-Chinese, Peking, October 19, 1972; NCNA, Peking, October 3, 1973; NCNA-Chinese, Peking, January 3, 1974; Hsiang Jung, op. cit.

29. JMJP, August 5, 1964, p. 1.

30. Ta Kung Pao, Peking, January 21, 1964, p. 1.

31. JMJP, July 26, 1965, p. 2.

32. Ibid., April 29, 1965, p. 2.

33. Hsin-hua Pan-yueh-k'an [New China Semimonthly], Peking, No. 4, 1958, pp. 97-98.

34. Arika Uchimi, "Problems in Deferred Payment to China," Ajia keizai jumpo [Asian Economy Thrice-Monthly], Tokyo, June 1964, pp. 11-14 (translated by the Joint Publications Research Service, JPRS No. 26,466 [Washington, D.C., September 21, 1964], pp. 13-18).

35. JMJP, November 25, 1957, p. 3.

36. "Chemical Fertilizer Industry Energetically Supports Agriculture," Ta Kung Pao, Hong Kong, October 1, 1962, p. 15.

37. NCNA, Peking, February 13, 1963.

38. JMJP, December 10, 1962, p. 2; June 11, 1963, p. 1; NCNA, Peking, September 27, 1963.

39. China Reconstructs, Peking, Vol. XIV, No. 6 (June 1965), p. 30.

40. "China's Nitrogen Fertilizer Industry Steps into Top World Ranks," China Reconstructs, Peking, Vol. XVIII, No. 5 (1969), pp. 11-15.

41. NCNA-English, Peking, December 30, 1968.

42. NCNA, Sian, May 19, 1969.

43. Ibid., Nanking, August 27, 1961.

44. Hsin-hua Pan-yueh-k'an, No. 4, 1958, pp. 97-98.

45. Li Ch'ing-k'uei, op. cit.

46. NCNA-English, Sining, January 3, 1966.

47. "China's Foreign Trade in 1969," Current Scene, Hong Kong, Vol. VIII, No. 16 (1970), pp. 1-8; "The Food and Population Balance: China's Modernization Dilemma," Current Scene, Vol. IX, No. 6 (1971), pp. 1-7.

48. Chinese Affairs Division, Ministry of Foreign Affairs, Tokyo, "The Present State of Japan-China Trade," Current Scene, Vol. VIII, No. 9 (1970).

49. KYODO, Tokyo, dispatch in English, March 1, 1971.

50. YOMIURI, Tokyo, June 21, 1969; Food and Agriculture Organization of the United Nations, The State of Food and Agriculture, 1967 (Rome, 1967), p. 66.

51. Chang K'ai-yen, "A Discussion of Certain Questions on the Promotion and Use of Phosphorus Fertilizer," JMJP, April 17, 1962, p. 5.

52. Kung-jen Jih-pao [Daily Worker], Peking, September 13, 1962, p. 1.

53. JMJP, October 20, 1962, p. 2.

54. Kung-jen Jih-pao, February 20, 1963, p. 1.

55. Chin Yen, "Pay Attention to Reduce the Damages and Losses of Chemical Fertilizers in Transit," Nan-fang Jih-pao [South China Daily], March 20, 1963, p. 2.

56. Snow, op. cit.

CHAPTER 13

SOIL CONSERVATION

Through many centuries Chinese farmers have maintained the productivity of the land by means of terracing, fallowing, manuring, crop rotation, and other soil conservation devices; but there are extensive areas where land has declined in productivity because of lack or inadequacy of drainage, irrigation, fertilization, or proper management. According to an estimate made in 1958, approximately 600 million mou (100 million acres), or more than one-third, of the cultivated land in China were low productivity—saline, alkaline, acidic, sandy, swampy, or erosive.[1]

The improvement and conservation of soil thus commands great importance in the country's agricultural development. The solution to this problem is to be found in the development of a program to promote better use of land by the adoption of practices designed to reduce water runoff. To this end the Communist government has given considerable attention to the assessment, utilization, and preservation of soil resources, including soil surveys and various means for controlling soil erosion and improving land fertility.

SOIL SURVEYS

A soil survey includes classification, mapping, and study of the best uses to which the soil can be put. Such a study supplies information needed in highway construction and for military purposes, but its most important use is in the development of agriculture. It is an essential guide to research programs for crops, soils, and fertilizers, and for planning the best use of land. The information that it supplies may prevent serious mistakes, fruitless efforts, and a waste of soil, forest, and other natural resources.

A program for the assessment of soil resources was begun in China in the 1930s by the newly organized Soil Survey Section of the National Geological Survey, with the assistance of several American soil experts. Charles F. Shaw, a soil scientist of the University of California who served as visiting professor at the University of Nanking in the spring and summer of 1930, was assigned to train field enumerators in the methods of obtaining soil samples from 16,786 farms in 168 localities of 22 provinces for the study of land utilization in China (1929-33). At Shaw's recommendation the Soil Survey Section was created at the National Geological Survey, and Robert L. Pendleton of the University of the Philippines, who had previous extensive experience in soil survey in India, was appointed for a two-year period (from the spring of 1931 to the spring of 1933) to continue the training of the land utilization investigators at Nanking. Pendleton, in collaboration with R. L. Chang, L. C. Chen, and K. C. Hou, also made intensive soil surveys of the Salach'i area in Suiyuan Province and the Tat'ung area in Shansi Province. In addition he conducted a reconnaissance soil survey of a portion of Kwangtung Province.

During the period from the spring of 1933 to the spring of 1936, W. H. Wong, director of the National Geological Survey, employed James Thorp of the U.S. Department of Agriculture to make an extensive soil reconnaissance survey of China. His findings and appraisal were published in 1936 by the National Geological Survey under the title Geography of the Soil of China. He also wrote a summary as a chapter on soil in J. Lossing Buck's Land Utilization in China (Nanking, 1937). By the end of 1936, the National Geological Survey had published 14 soil bulletins. Eight other intensive local soil studies, all by Chinese soil scientists, were published by Sun Yat-sen University (Canton, 1932-36).[2]

The soil survey work was interrupted by the Sino-Japanese War and was resumed by the Research Institute of Soil Science, established at Nanking in 1952 under the Chinese Academy of Sciences. For the first few years the institute's principal work was the preparation of a new system for a nationwide soil survey, for which Li Ch'ing-k'uei, president of the Chinese Society of Soil Science, gave the following explanation:

> Between the 1930s and the 1950s, China's soil classification was primarily based on a combination of the systems designed by the world's two leading soil scientists, V. V. Dokuchaev of Russia and C. F. Marbut of the United States. For the major soil types, American and European continental names were adopted with the supplement of some Chinese names used by local farmers. In preparing regional or local soil maps, the types of the

soils were named after the localities where the soils were first found. A major shortcoming of this system is that it hardly reflects the effect of cultivation on the fertility of each type of soil. In view of this shortcoming, another system has to be adopted.[3]

The national soil survey, covering approximately 1.6 billion mou (267 million acres) of farmland, began in late 1958, following a conference of soil scientists held at Hsinhsing, Kwangtung, in October of that year under the joint auspices of the Chinese Academy of Agricultural Science and the General Office of Land Utilization of the Ministry of Agriculture. The survey, using a combination of native and Western methods, was conducted under the supervision of the CCP committees at different levels in all localities. Its personnel was composed mainly of some 7 million peasants who were organized into groups, each headed by a technical staff. Emphasis was laid on investigating the changes in the soil resulting from the methods of cultivation of various crops under different natural conditions and in different areas. According to Vice-Minister of Agriculture Liu Jui-lung, the survey in effect summed up the experience of Chinese peasants in handling the soil over thousands of years. It provided data for plowing, cultivating, fertilizing, and irrigating different types of soil and for improving the country's soil in a planned way.[4]

The survey, including the classification of the soils, was completed in a preliminary form during the winter of 1959-60. In the opinion of Ch'eng Chao-hsüan, vice-president of the Chinese Academy of Agricultural Science, it laid a scientific foundation for the practice of planting crops according to the nature of the soil, for the practice of rational crop rotation, and for the improvement of soil by the communes.[5] The survey also formed the basis for a general soil map of China, regional soil maps, a map showing China's soil fertility, a map showing the present use of land, and a book on the experience of Chinese peasants in identifying, utilizing, and improving soils. These publications were reported to be in preparation in early 1962.[6]

The nature of the national soil survey may be illustrated by the provincial survey in Kwangtung, which covered all the cultivated land in the province's 120 hsien and cities. After six months of fieldwork in which nearly 180,000 technicians and peasants participated, soil maps and land utilization maps of 1:10,000 or 1:5,000 scale were prepared for all communes, and on a smaller scale for various special districts and hsien. A 1:200,000 scale soil map for Kwangtung and a publication on the soils of the province have been completed.[7]

Another example is the soil survey conducted in the suburban areas of Peking in the winter of 1958-59. Under the direction of a soil research group of the Chinese Academy of Sciences, the Peking

Municipal Academy of Agricultural Science, and the Peking Municipal Council, the survey included the participation of 39 students from the department of geography of the Peking Normal University, more than 2,000 peasants, 200 cadres, and a number of party workers who had been dispatched to the rural areas for agricultural work training. The whole area under survey was divided into eight districts, each under the direction of an expert from the Chinese Academy of Sciences. Maps of 1:10,000 scale were prepared, showing the types and distribution of high-yield farmland, measures taken to improve land, distribution of canals, and other features in each commune or district.

According to Hsiung I, leader of the Soil Research Group of the Chinese Academy of Sciences, one of the problems in conducting the Peking soil survey was how to select and unify the names of various types of soil that traditionally had been used by the peasants. Many of the names sound very lively and are suggestive of the characters of the soil, but at the same time they are rather misleading and not suitable for use in the classification of soils.[8]

Other important surveys in connection with soil and water conservation conducted since 1949 included those for the areas along the Yangtze, Yellow, Huai, Heilung, Liao, and other rivers, and those for the subtropical areas of south China. During the First Five-Year Plan (1953-57) a soil survey of 640 million mou (107 million acres) of wasteland throughout the country was conducted and a total of 800 million mou (133,334,000 acres) of wasteland were surveyed for soil and water conservation purposes.[9]

One of the most important studies on the soils of China was made by a Russian agronomist, V. A. Kovda, in 1954-57. Entitled Ocherki prirody i pochv Kitaya (Soils and Natural Environment of China), the book is based on materials compiled by the Chinese Academy of Sciences, results of the field studies conducted by the writer in collaboration with Chinese scientists, and findings in the laboratories of the Soil Research Institute of the Soviet Academy of Sciences. The publication, which includes a map of 1:10,000,000 scale showing the geographic distribution of 64 soil types, was published in 1959 by the Soviet Academy of Sciences Press in Moscow.[10]

Despite continued efforts to improve soil science in the PRC, there remain widely different opinions on soil classification and nomenclature among Chinese scientists. The problem was discussed at length at the annual meeting of the China Society of Soil Science held in Shenyang on August 15-25, 1963, during which a provisional system of soil classification based on the findings of Chinese scientists and the experiences of Chinese peasants" was adopted.[11] But thus far no important development in this connection has been reported.

IMPROVEMENT OF LAND OF LOW FERTILITY

As already stated, approximately 600 million mou, or more than one-third of China's cultivated land, are of low productivity. The NPAD urged the APCs (now the communes) and the state farms to actively improve and utilize saline and alkaline land, red-soil land, low-lying land, sandy land, and poor land of other types; to guard against the salification and alkalization of land; and to terrace hilly land energetically and in a planned manner (Article 12).

A directive issued by the Central Committee of the CCP and the State Council on August 29, 1958, stated that of the 470 million mou (78.3 million acres) of saline and alkaline land, red-soil land, sandy land, muddy fields, and other land of low fertility (excluding swamps) throughout the country, more than 140 million mou (23.3 million acres) had been improved by that date. It urged the communes to improve the remaining 330 million mou (55 million acres) within two or three years.[12] Toward the end of 1959, more than one-half of the 600 million mou (100 million acres) of land of low fertility had been improved by means of terracing and other processes of soil conservation, according to Minister of Agriculture Liao Lu-yen.[13] Another account, given by T'an Chen-lin, vice-premier of the State Council, revealed that of the more than 700 million mou (116.7 million acres) of lowland subject to waterlogging and low-yielding land of various types, "initial improvement" had been effected on 450 million mou (75 million acres), or 60 percent of the total.[14]

Late in 1975 NCNA reported that during the first quarter-century of the PRC, more than half of the alkali land in north China, 8 million hectares of red earth in ten provinces and one autonomous region in the middle and lower reaches of the Yangtze River, and one-fifth of the erosion-impoverished land had been improved. In addition more than two-thirds of the land susceptible to waterlogging had been protected by drainage facilities; and poorly sloping fields and cold, water-soaked land in mountainous areas had been improved. Forest belts totaling several thousand kilometers and extensive water conservation projects had brought about an increased number of green oases in the desert areas of northwest and north China.[15]

According to an estimate made by the Research Institute of Water Conservation and Hydroelectric Power, China has some 300 million mou (50 million acres) of saline and alkaline land, of which about 100 million mou are under cultivation. Most of this land is located in the areas where the rate of evaporation exceeds that of rainfall, including the plains of northwest, northeast, and north China, basins in the interior, areas near Po Hai Bay, and the coastal areas in northern

Kiangsu. In north China saline and alkaline land represents about 10 percent of the cultivated land and about 20 percent of the irrigated land.[16]

Salification and alkalization of land result from improper farming practices, particularly improper irrigation and drainage. The peasants in north China had generations of experience in preventing and controlling salification and alkalization. Their measures included the leading of floodwater to wash away salt, the drainage of land to grow rice, the building of embankments to stabilize planting, and the storage of fresh water to suppress salt. On the plain of Hopei, which was always on the margin of salinity, the farmers kept the land fertile by carefully constructing small dams and elevated paths around the fields to prevent the mounting of the underground salt and by using organic fertilizer and intensive cultivation to block the capillaries that bring up the salt water and to reduce evaporation.[17]

These measures are effective only when they are carried out together with the traditional methods of farming. When the communes were set up in 1958, many of these old ways were abandoned in favor of new ways that have proved less effective; inadequate drainage in particular has caused the level of the groundwater to rise, bringing more salt to the surface. It was claimed that during the first decade of the Communist regime, some 20 million mou of saline and alkaline land were improved.[18] But improper water conservation measures, particularly those taken during the Leap Forward campaign, apparently have resulted in the salification and alkalization of land in many areas (see Chapter 6).

Red soils occupy 75 percent of all land in Kweichow, 70 percent in Hunan, 56 percent in Kwangtung, 55 percent in Kwangsi, 46 percent in Kiangsi, and 40 percent in both Chekiang and Fukien.[19] In 1950-61 the Kiangsi Agricultural Research Institute made a series of studies on the possibilities for improving the red-soil (lateritic soil) land already in use and for reclaiming some of the red-soil wasteland in Kiangsi Province. After examining the methods used by the peasants in growing rye, peanuts, sweet potatoes, and other principal crops, the institute recommended that proper amounts of farmyard manure be applied to red soil in addition to chemical fertilizers and that certain technical measures used by local farmers be adopted. It was reported that by the end of 1961, more than two million mou of red-soil wasteland were reclaimed and several million mou of red-soil farmland were improved in Kiangsi Province.[20] In several other provinces vast tracts of red-soil land were reported to have been improved by applications of farmyard manure and lime to increase the organic content and reduce the acidity of soil.[21]

According to incomplete statistics for ten provinces and one autonomous region in south China (Hunan, Kiangsi, Fukien, Yunnan,

Kweichow, Kwangtung, Kwangsi, Szechuan, Chekiang, Anhwei, and Hupeh), more than 120 million mou (20 million acres) of land with red loam soil had been improved and utilized by the end of 1974. The total acreage of red loam soil accounts for approximately 41.7 percent of the total land area of the provinces and autonomous region. About 150 million mou (25 million acres) of red loam soil had been turned into paddy fields, more than 100 million mou (17 million acres) had yet to be developed, and more than 200 million mou (34 million acres) were barren mountains.[22]

China's low-lying agricultural land is distributed mainly along the rivers and seacoasts in the provinces of Hopei, Honan, Shantung, Shansi, Kiangsu, Anhwei, Hunan, Hupeh, Kiangsi, Kwangtung, Liaoning, Kirin, and Heilungkiang and the municipalities of Peking and Tientsin. These areas are important in producing wheat, cotton, soybeans, and other principal crops; but they are subject to frequent floods and waterlogging. Even in normal years an average of 90-97 million mou (15-16 million acres), or 9-10 percent, of the cultivated land in the low-lying areas was flooded or waterlogged. In years when the precipitation was extraordinarily heavy, as in 1954-56, the acreage affected reached 240 million mou (40 million acres), or 22 percent, of the cultivated land in these provinces and municipalities.

In December 1957 a conference was called in Yangliuch'ing, Hopei, by the Ministry of Agriculture and the Ministry of Water Conservation to discuss measures for improving the low-lying agricultural land. The projects that had been carried out in the Tientsin Special District, the Ts'anghsien Special District, and a few other localities were used as models. The conference participants generally agreed that the peasants and local governments should play an increasingly important role in carrying out various improvement measures because the central authorities alone could not pay the heavy costs involved.[23]

In recent years the "raised fields" method of reducing waterlogging and salinity in low-lying areas, a method evolved by peasants in north China, has been popularized in the basin of the lower reaches of the Yellow and Huai rivers. The raised fields consist of rectangular farm plots bordered on two or three sides by deep drainage ditches. Over large tracts of land, the earth dug out from the drainage ditches is used to raise the height of the fields and cover the saline topsoil. Such fields prevent the accumulation of surface water after heavy rainfall in the low-lying areas and also reduce salinity.[24]

In 1975 more than two-thirds of China's 23.3 million hectares of low-lying farmland were reported to have been provided with drainage systems.[25]

PROTECTION AGAINST SOIL EROSION

Approximately 2.4 million square kilometers (600 million acres), or roughly one-fourth of China's total land area, are subject to soil erosion of all kinds; and surface erosion and gully erosion are apparent over approximately 1.5 million square kilometers, or one-sixth of the total land area. The areas most affected include the basin of the Yellow River, covering about 600,000 square kilometers, especially within the huge loop of the river; the basin of the Yangtze River, covering 160,000 square kilometers; an area surrounding Peking and in the mountains to the northeast of the bend of the Yellow River, totaling 130,000 square kilometers; and in the basin of the Huai River, covering 75,000 square kilometers.

Among the climatic factors causing soil erosion are the frequent and violent sandstorms and the alternation of frost and thaw. Another cause is the varied topography, which is very mountainous in most parts of the country. Geologically the situation is aggravated by the abundance of sedimentary rocks (marine facies of sandstone, clay, and sands resulting in soft, hilly relief). The deep, recent Quaternary deposits—such as moraines, superficial glacial deposits, river gravels, alluvium, and loess—furnish inexhaustible matter for erosion. Finally, erosion has been accelerated by the destruction of forest and the formation of torrents as a result of improper clearing, grazing, and burning.[26]

By the end of 1959, preliminary measures for water and soil conservation had been carried out on 890 million mou (150 million acres), or 40 percent of the total of 2,224.5 million mou (370.5 million acres) of land seriously affected by soil erosion.[27] As already mentioned, one of the areas where soil erosion has been most serious is the area along the middle reaches of the Yellow River in northwest China. This is also one of the biggest loess plateaus in the world. It stretches from the Ch'ilien Mountains in the west to the T'aihang Mountains in the east and from the Tach'ing Mountains in the north to the Chinling Mountains in the south. It traverses Chinghai, Kansu, Ninghsia, Inner Mongolia, Shensi, Shansi, and Honan, and covers a total area of about 600,000 square kilometers. On the loess plateau there are about 26,000 square kilometers of gullies.

The Yellow River flows from the western part of the plateau eastward, down the high mountains and giant valleys. On the way it is joined by several score of principal tributaries that bring with them an abundance of water from all directions. For centuries agricultural production in the area has been jeopardized by the loss of water and soil on the hill slopes and gullies, and the water and soil thus lost have greatly reduced the acreage of arable land on the middle reaches of the Yellow River. Moreover, because the riverbed on the middle reaches is steep, the greater part of the silt carried away by the currents is deposited on the lower reaches.

It is estimated that approximately 400 million tons of silt are deposited on the lower reaches of the Yellow River every year, raising the riverbed and reducing the capacity of the river course year after year. This has resulted in frequent floods, especially after the downpours during the months from June to September, when 70 to 80 percent of the average annual precipitation of 300-500 millimeters is recorded. The situation has been worsened by the fact that less than 5 percent of the loess is covered by vegetation.[28]

In the late 1960s, particularly during the slack seasons, the PRC launched a number of campaigns for the control of soil erosion on the middle reaches of the Yellow River. Such projects as the construction of terraces, check dams, ditches, and field ridges across gullies to capture rainwater, and the planting of trees and grass on steep slopes, have been encouraged. Many communes participated in the campaigns by dynamiting rocks and making bricks and tiles in their own kilns to provide building materials for check dams and protective walls around the terraced fields.

By the end of 1964, 20 experiment stations and 51 technical aid stations had been set up in the loess area to study water and soil conservation problems and to extend erosion control experience to the communes.[29] A year later some 160,000 hectares of land had been terraced, turned into strip-cropping fields, or brought under irrigation by more than 100 hsien marked out as key centers for antierosion work. Trees and grass had been planted on 52,000 hectares,[30] and a number of water and soil conservation stations and demonstration farms had been established to strengthen the antierosion work.[31]

Despite these efforts, water and soil conservation work on the middle reaches of the Yellow River has been sporadic, inadequate, and uncoordinated. Moreover, human destruction of the soil has not yet ended, and loss of soil is appearing in new areas. Since in many places communicating paths have not been built between the terraces, peasants who carry fertilizer break through the barriers and therefore open the way for erosion. Afforestation in such areas is difficult because of the lack of rain. The ineffectiveness of years of afforestation in the past also was due to the improper selection of types of trees and the destruction of young trees by men and animals.[32]

AFFORESTATION

In 1971 China's forested area was about 111.8 million hectares (1,677 million mou), or only 12 percent of the total land area, as compared with 41 percent in the Soviet Union (1973), 32 percent in the

United States (1969), 20 percent in India (1972), and 69 percent in Japan (1972).[33] It is clear that China is inadequately forested. Several projects to reconstitute and improve natural forests and to create new ones were included in the First Five-Year Plan. These projects were expanded in 1956 into a 12-year (1953-65) program to "make China a green country" that envisaged the afforestation of 92 million hectares (227.2 million acres). The new forests, together with the 76.6 million hectares (189.2 million acres) of natural forests, would make up a forest domain of 168.6 million hectares (416.4 million acres) by 1965; and the proportion of forested land in China would then rise to 19 percent. The goal was not reached.

The most urgent objective of the program was to relieve the enormous timber shortage in China, which shackled the desired expansion of many industries, particularly building construction. The long-term significance of the program was to lessen the possibilities of soil erosion, floods, and droughts and to prevent the silting of rivers and hydroelectric reservoirs. The program also was intended to improve hygiene and public health and to beautify urban and suburban areas.[34]

The APCs (now the communes) have been relied on to carry out the bulk of the afforestation program. Of the 497.86 million mou (82,977,000 acres) of land afforested in 1950-58, 21.7 million mou (3,617,000 acres), or about 4 percent, were afforested by some 2,000 state forest farms, while the remaining 96 percent were afforested by the communes.[35] In 1964 the communes had more than 30,000 forest farms and forest teams in operation, with some 200,000 commune members engaged in the afforestation work. By the end of that year, more than 3,000 state forest farms (six times as many as in 1957) had been established, employing 200,000 workers.[36] In 1972 the number of state forest farms exceeded 3,700.[37]

While the communes are counted on to plant trees in the villages, the state forest farms are mainly responsible for undertaking the planting of large tracts of timber forests, shelter belts for water and soil conservation, windbreaks, sandbreaks, and shelter belts along the seacoasts. State efforts also have been directed toward the reforestation of degraded natural forests and the afforestation of bared hills and steppe land unsuitable for agricultural uses. The state also has the right to take over all uncultivated land, where reforestation operations are carried out as a public service to protect land against erosion. The communes, on the other hand, are encouraged to participate in tree-planting work in accordance with the national programs and plans, with financial and technical assistance from the state.

Table 13.1 shows that early efforts to develop the forests did not yield significant results but a considerable increase in forest acreage was reported during the Leap Forward campaign. In 1958 and 1959

TABLE 13.1

Afforested Area in the PRC, 1950-59
(thousand mou)

Year	Shelter Belts	Timber Forests	Other	Total
1950	1,010	210	680	1,900
1951	3,790	1,150	1,820	6,760
1952	8,540	3,310	4,430	16,280
1953	6,250	6,710	3,730	16,690
1954	5,080	9,540	2,870	17,490
1955	5,900	14,210	5,550	25,660
1956	20,270	36,810	28,770	85,850
1957	14,920	26,020	24,390	65,330
1958	50,020	90,230	121,650	261,900
1959	n.a.	n.a.	n.a.	280,000

Note: n.a. = not available.

Sources: State Statistical Bureau, Ten Great Years (English ed.; Peking, 1960), p. 133; JMJP, January 23, 1960, p. 1.

approximately 542 million mou of land were afforested. But in 1964 China's forest area still constituted only about 10 percent of the country's total land area.[38] Thus progress was much slower than that anticipated in the 1953-65 afforestation program, which called for 19 percent of China's total land area to be forested by 1965. The Ministry of Forestry attributed the slow progress to the poor quality of the afforestation work as a whole; the low rate of forest regeneration, which did not match the rate of felling; and the serious destruction of forests caused by fire and improper lumbering.[39]

The "Regulations Governing the Protection of Forests," promulgated by the State Council on May 27, 1963, were regarded by the Chinese press as having an important bearing on the country's industrial and agricultural production as well as on the people's daily life. While acclaiming the government's work of afforestation and protection of forests during the first decade of the Communist regime, the People's Daily stated in an editorial that many localities still lacked experience in this field, that "quite a few" localities failed to pay proper attention to it, and that therefore the work of afforestation and protection should be intensified.[40] In a notice issued in May 1965, the Ministry of Forestry again called on the country to "genuinely" implement the May 1963 regulations, and in particular to strengthen field management of saplings and young trees in order to prepare sufficient supplies for use in future large-scale afforestation programs.[41]

Terracing remains one of the most important methods used in China for controlling soil erosion as well as water supply. On the loess plateau in northwest China particularly, land is so rough that extensive irrigation is impossible. For centuries the Chinese farmers have met the situation with terracing. Stone-faced terraces and bench terraces are used to hold as much water as possible. In some areas of Shansi Province, the terraces are graded and sloped toward the mountain to hold rainfall. At the center of each terrace is a carefully constructed stone-faced spillway to carry off any excess water.

In the rice-producing areas terracing is even more extensive because rice fields must be level to hold water. This leveling with plows, shovels, and a special pronged iron leveler drawn by a water buffalo remains one of the greatest aids to the conservation of both soil and water in the PRC today. During the winter of 1957-58 more than 51 million mou of terraces were built. This was equivalent to 26 percent of the total area of hilly cultivated land.

MULTIPLE CROPPING

Multiple cropping, or growing more than one crop per unit of land per year, is highly developed in some countries, particularly in East Asia, as a means of increasing agricultural production. The methods of multiple cropping are based on such local natural and economic conditions as climate, soil fertility, topography, and supply of water, fertilizers, and manpower. In China different systems of multiple cropping are found in different areas within a province. The process is not a simple matter of sowing more than one crop per year. Moreover, multiple cropping requires substantially greater inputs of capital and labor and more scientific management.

The degree of multiple cropping is usually represented by the "multiple-sown area index," also called "multiple-cropping index," which is derived by the following calculation: sown area divided by cultivated area multiplied by 100. The targets provided in the NPAD were to raise the average multiple-sown area index (including green manure crops) as follows in the 12 years beginning in 1956: (1) in areas south of the Wuling Mountains, 230 percent; (2) in areas north of the Wuling Mountains and south of the Yangtze River, 200 percent; (3) in areas north of the Yangtze River and south of the Yellow River, the Chinling Mountains, and the Pailung River, 160 percent; (4) in areas north of the Yellow River, the Chinling Mountains, and the Pailung River and south of the Great Wall, 120 percent; and (5) in areas north of the Great Wall, in general all cultivated land should be fully utilized, the area of unworked land should be reduced, and, wherever possible, energetic steps should be taken to expand the multiple-cropping areas (Article 9).

TABLE 13.2

Natural Conditions of the Five Regions and Proposed Increases of Multiple-Sown Area Indexes, 1956-67

Region	Annual Precipitation (millimeters)	Average Annual Temperature (Centigrade)	Number of Frost-Free Days Each Year	Proposed Increase of Multiple-Sown Area Index, 1956-67
South of the Wuling Mountains	1,500-2,000	20-24°	320	184-230
North of the Wuling Mountains and south of the Yangtze River	1,250-1,750	16-20°	250-300	154-200
North of the Yangtze River and south of the Yellow River, the Chinling Mountains, and the Pailung River	500-1,000	14-17°	200-250	156-160
North of the Yellow River, the Chinling Mountains, and the Pailung River and south of the Great Wall	300-500	8-15°	150-220	113-120
North of the Great Wall	200-750	0-8°	120-170	"as much as possible"

Source: Nung-yeh K'o-hsueh T'ung-hsun [Agricultural Science Bulletin], Peking, No. 1, 1958, p. 1.

TABLE 13.3

Cultivated Area, Sown Area, and Multiple-Sown Area Index, in the PRC, 1949-58

Year	Cultivated Area (thousand mou)	Sown Area (thousand mou)	Multiple-Sown Area Index
1949	1,468,220	n.a.	n.a.
1950	1,505,340	n.a.	n.a.
1951	1,555,070	n.a.	n.a.
1952	1,618,780	2,118,840	130.9
1953	1,627,930	2,160,530	132.7
1954	1,640,320	2,218,880	135.3
1955	1,652,350	2,266,220	137.2
1956	1,677,370	2,387,590	142.3
1957	1,677,450	2,358,660	140.6
1958	1,616,800	2,344,020	145.0

Note: n.a. = not available.

Source: State Statistical Bureau, Ten Great Years (English ed.; Peking, 1960), p. 128.

Like the proposed boundaries dividing the country into three regions for yield targets, the boundaries dividing the country into five regions for the purpose of multiple-sown area index were drawn on a somewhat arbitrary basis. Why the Wuling Mountains (the five principal mountain ranges traversing Kiangsi, Kwangtung, Hupeh, Hunan, and Kwangsi) and the Pailung, a small river in southern Kansu, were used for the demarcation of the regions was not explained, although their natural conditions and the proposed multiple-sown area indexes were briefly indicated (see Table 13.2).

As shown in Table 13.3, the total cultivated area for China as a whole increased steadily from 1949 to 1957; and the total sown area increased steadily from 1952 to 1956 and then declined in 1957 and again in 1958. The resulting annual multiple-sown area index rose from 130.9 in 1952 to 145.0 in 1958. Even the highest figure during 1952-58 was still lower than the index of 149.0 given in J. Lossing Buck's study of 16,786 farms in 22 provinces for the period 1929-33.[42] Buck's index might have been somewhat lower if all of Mainland China had been included in his study. Even so, the figures given by the PRC do not indicate much success in multiple cropping.

In their efforts to promote multiple cropping, PRC authorities have changed crop cultivation systems in some areas without consideration of local conditions. In many cases the changes have produced bad results. In the greater part of the Yellow River valley, for instance, the three-crops-in-two-years system was traditionally

used. In the 1950s and 1960s the system was changed to two crops in one year. Although this change benefited some localities with improved soil fertility, water conservation, and manpower, it produced undesirable effects on the agricultural production in other areas, where economic and natural conditions remain unchanged. A survey of the Shih Li P'u production brigade of the Fa Hsin Commune in Shun-i hsien near Peking revealed that the three-crops-in-two-years system (spring corn in the first year and winter wheat and summer corn in the second year) produced an average yearly yield of 400 chin per mou. After the change to the two-crops-in-one-year system (wheat and summer corn), the yearly yield averaged only 336 chin per mou, a decrease of 19 percent from the former system. The survey also showed that under the three-crops-in-two-years system, the production cost was 32 percent lower, the total value of production was 9 percent higher, and the net income was 115 percent higher than under the two-crops-in-one-year system.[43]

An effort also was made to extend the area devoted to the production of two rice crops per year in the basins of the Yangtze River. Two-crop rice culture had long been practiced in parts of southern China where climate and other factors permitted. Its increase was particularly rapid in Szechuan Province, from 80,000 mou (13,300 acres) in 1955 to 4.5 million mou (750,000 acres) in 1956 and 8.97 million mou (1,495,000 acres) in 1957. Two-crop rice culture was attempted with varying levels of success in several areas between 32 and 33 degrees north, including Kuangyuan in northern Szechuan, Ank'ang in southern Shensi, Kuch'eng in northern Hupeh, Hsinyang in southern Honan, Pengfu in Anhwei, Lihsiaho in northern Kiangsu, and the Yenmen section of Kiangyu hsien in Szechuan, which has an elevation of 800 meters above sea level.[44]

As a whole the effort to extend the two-crop rice culture area was not successful, since low yields and uncertain harvests still existed for the late crop in many localities. In 1958 the late crop yielded only 150-250 chin per mou in various provinces, compared with 400-500 chin per mou for the early crop. For several years hundreds of thousands of mou of the late crop had bad harvests or none at all. In addition to inadequate water supply and fertilization, another reason for the low yield of the late crop is that there are few late-crop varieties of high productivity and high quality; the grain tends to shatter easily. Still another reason is that the time of harvesting of the first crop and planting of the second crop comes too late in the season.[45]

EXPANDING THE AREA SOWN TO HIGH-YIELDING CROPS

Rice, corn, and potatoes are the most important high-yielding crops in the PRC. The NPAD called for the expansion of the area sown to rice by 250 million mou in the 12 years beginning in 1956. It also encouraged the development of corn, potatoes, and other high-yielding crops according to the local needs and dietary habits (Article 10).

The average yields, in chin per mou of sown area, of the principal grain crops for 1949-55 were given by Wang Shou, director of the Research Institute of Crop Breeding and Cultivation of the Chinese Academy of Agricultural Science, as follows:

Rice	312.2	
Potatoes	231.3	(converted into grain-equivalent at the ratio of four chin to one)
Corn	179.6	
Millet	136.5	
Kaoliang*	133.5	
Soybeans	102.0	
Wheat	99.7	

Wang Shou indicated that under the program to promote the growth of high-yielding crops, the sown area for rice in 1956 increased by 15.9 million mou over 1955; that for corn increased by 48 million mou; and that for potatoes increased by 13 million mou. As a result of the increase in sown areas for these high-yielding crops and the adoption of other measures, in 1956 China's total grain production increased by more than 15,000 million chin over that in 1955, despite the serious natural calamities in 1956. Wang Shou further revealed that in 1956 the total sown area for rice, corn, and potatoes constituted only 49.7 percent of the total sown area for all grain crops, yet the total production of these three crops amounted to 70.8 percent of all grain production in the PRC. For rice the total sown area constituted only 26.7 percent of that for all grain crops, yet the total production of rice was 45.5 percent of that for all grain crops. From these facts Wang visualized the great possibilities for further expanding the sown areas for high-yielding crops. However, he pointed out that in 1956 many attempts to expand sown areas for corn and potatoes, without consideration of local needs and dietary habits, caused serious waste as well as transportation and storage difficulties. Also, the use of seeds of certain varieties not suitable to local climate and soil conditions resulted in low production of high-yielding crops, as in the case of the "Golden Queen" variety of corn.[46]

In the opinion of J. Lossing Buck, the expansion of the area sown to potatoes in the PRC is a good practice; but the attempt to

*A kind of grain sorghum.

increase the area sown in rice may have resulted in displacing other, more profitable crops. In general, rice cultivation in China before World War II extended too far up hilly land, where the water supply was not dependable. Farmers in Szechuan called it "gambling with heaven"; and some of them discontinued growing rice on hilly land in favor of corn and soybeans, which gave higher yield.[47]

DEEP PLOWING

On August 29, 1958, the Central Committee of the CCP and the State Council jointly issued a directive urging the communes to adopt deep plowing as a measure to increase agricultural production. The directive stated that all suitable land should be deep-plowed once every three years. The depth of plowing generally should reach more than one ch'ih (14 inches), and the highly productive fields should be plowed to a depth of at least two ch'ih. In the areas with "practical difficulties," deep plowing should be carried out in two stages: first to six or seven ts'un (9.5 or 10.0 inches) and later to more than one ch'ih.[48]

It was estimated that by the end of 1958, more than 800 million mou (133.5 million acres) of land, or nearly half the total area of cultivated land, had been deep-plowed. Deep plowing was said to have improved the quality of the soil, prevented plants from bending and being uprooted, promoted microbiological activities, and aided in weed control and the prevention of certain plant diseases and pests.[49]

According to Ch'eng Chao-hsuan, vice-president of the Chinese Academy of Agricultural Science, surveys and experiments proved the favorable effect of deep plowing on agricultural production, and the effect of one course of deep plowing might last for about three years.[50] However, others maintain that deep plowing, like close planting, has its limitations and can contribute to agricultural production only under certain circumstances. They argue that the depth of plowing should be based on the type of soil and the kind of crop grown, among other factors, and that deep plowing should be adopted in conjunction with other measures for improving agricultural production, particularly water conservation and fertilization.[51]

During the Leap Forward campaign explosives were used in many areas where deep plowing was practiced, in order to save labor. The communes were urged to develop explosives made of native material. A technical group in the Research Institute of Mechanical Science of the Chinese Academy of Sciences conducted a series of experiments on explosives made of low-cost chemicals, principally potassium nitrate (75 percent), sulfur (10 percent), and charcoal (15 percent). The use of explosives in cultivating land was said to be less subject to topographical limitations and to make it possible to loosen land with less labor and at higher speed.[52] However, some serious questions remain unanswered. Does the loosening of soil by means of explosion meet the requirements for crop growth? Does the use of explosives ruin the composition of soils and thus affect the growth of crops? In any case it would appear that the experimental use of explosives in deep plowing has not become popular.

THE PROSPECTS

The soil conservation program in the PRC consists mainly of mass campaigns, with the communes and their constituents playing a much more important role than the central government. Peasants as well as cadres and technicians have participated in a great variety of activities, including such highly technical projects as soil surveys and soil classification. The collective afforestation by the communes has made a much greater contribution than the state forest farms in the program to "make China a green country." The local governments are also responsible for carrying out work to improve land of low fertility; to construct terraces, field border ridges, and silt-detention dams; and to apply farmyard manure, especially during the slack winter season.

Because of the large amount of labor needed for the soil conservation program, as well as for many other programs, there has been a constant labor shortage. One method of reducing the shortage would be to use more machines and to improve implements and tools; however, this approach has not been used successfully (see Chapter 17).

As in other aspects of the technical transformation of agriculture, much of the effort to conserve and improve soil fertility has been wasted as a result of improper planning and management. Moreover, in order to insure the success of certain key projects, such as soil erosion control along the middle reaches of the Yellow River, the central government would have to provide tremendous amounts of funds, machinery, and equipment, in addition to the contributions from the local units.

NOTES

1. Liao Lu-yen, "Ten Years of Glorious Accomplishments on the Agricultural Front," Nung-yeh K'o-hsueh T'ung-hsun [Agricultural Science Bulletin], Peking, No. 19, 1959, pp. 652-58; "Improving Farmland of Low Productivity," editorial, Chung-kuo Nung-pao [Chinese Agricultural Bulletin], Peking, No. 8, 1958, p. 21.

2. Information supplied to the author by J. Lossing Buck.

3. Li Ch'ing-k'uei and Yu T'ien-jen, "Several Questions Concerning China's Soil Science in Progress," K'o-hsueh T'ung-pao [Scientia], Peking, No. 2, 1964, pp. 118-26.

4. Liu' Jui-lung, "Let the Peasants Discover the Secret of Soils," Hung Ch'i [Red Flag], Peking, No. 3, 1959, pp. 26-30.

5. Ch'eng Chao-hsuan, "Basic Experience of High-Speed Development of Agricultural Science in China," Hung Ch'i, No. 16, 1960, pp. 24-30.

6. Ch'eng Chao-hsuan, "Great Leap Forward of Agricultural Production and Accomplishments of Agricultural Research," Nung-yeh K'o-hsueh T'ung-hsun, No. 19, 1959, pp. 659-61; NCNA, Peking, January 4, 1962.

7. Ch'eng Chao-hsuan, op. cit., pp. 659-60.

8. Hsiung I, "A Study of the Soil Classification System in the Soil Survey for the Suburban Areas of Peking," K'o-hsueh T'ung-pao [Scientia], Peking, No. 6, 1959, pp. 182-84.

9. Land Utilization Bureau, Ministry of Agriculture, "Great Achievements of Soil Work in China," Chung-kuo Nung-pao, No. 11, 1959, pp. 12-15.

10. An English translation of the book was issued by the Joint Publications Research Service, JPRS No. 5967 (Washington, D.C., October 31, 1960).

11. T'u-jan Hsueh-pao [Acta Pedologica Sinica], Peking, No. 3, 1963, pp. 341-42.

12. JMJP, September 11, 1958, p. 1.

13. Liao Lu-yen, op. cit.

14. T'an Chen-lin, "Strive for the Fulfillment, Ahead of Schedule, of the National Program for Agricultural Development," JMJP, April 7, 1960, p. 1.

15. NCNA-English, Peking, November 10, 1975.

16. Chang Tzu-lin and Huang Jung-han, "The Improvement of Saline and Alkaline Land and the Prevention of Salification and Alkalization in North China," Hung Ch'i, Nos. 15/16, 1962, pp. 47-56.

17. Hsiung I, "The Problem of Improving Saline and Alkaline Land in Hopei Province," Kuang-ming Jih-pao [Kuang-ming Daily], Peking, July 10, 1962, p. 2.

18. Chang Tzu-lin and Huang Jung-han, op. cit.

19. Lin Nung, "Poor Land Becomes Good Farms," Chi-hsieh-hua Nung-yeh [Mechanized Agriculture], Peking, No. 3, 1957, pp. 4-5.

20. Kuang-ming Jih-pao, January 19, 1962, p. 1.

21. NCNA, Peking, December 7, 1962; NCNA, Nanchang, September 15, 1963.

22. NCNA-Chinese, Peking, January 23, 1975.

23. Ho Chi-feng, "Great March Forward to the Low-Lying Land," Chung-kuo Nung-pao, No. 7, 1958, pp. 8-12; Li Ching-yu, "Improving Low-Lying Swampy Land and Converting Low Yield to High Yield," Chung-kuo Nung-pao, No. 7, 1958, pp. 13-17.

24. NCNA, Peking, November 16, 1965; NCNA-English, Peking, June 11, 1966.

25. NCNA-English, Peking, November 11, 1975.

26. Jean Messines, "Forest Rehabilitation and Soil Conservation in China," Unasylva, Rome, Vol. XII, No. 3, 1958, pp. 103-20.

27. T'an Chen-lin, op. cit.

28. Chao Ming-fu, "Let Us Change the Look of the Loess Plateau," K'o-hsueh Ta-chung [Popular Science], Peking, No. 6, 1963, pp. 1-3.

29. Chia Chen-lan, "An Inquiry into the Question of Building Steady and High-Yielding Farmland on the Northwest Loess Plateau," JMJP, May 19, 1964, p. 5; NCNA, Peking, December 3, 1964.

30. NCNA, Peking, January 9, 1966.

31. Peking Review, No. 28, 1965, p. 21.

32. Chang Sen, "Several Problems of Water and Soil Conservation on the Middle Reaches of the Yellow River," JMJP, June 4, 1963, p. 5.

33. Food and Agriculture Organization of the United Nations, Production Yearbook 1974, Vol. XXVIII-1 (Rome, 1975), pp. 3-8.

34. Messines, op. cit.; Food and Agriculture Organization of the United Nations, Production Yearbook 1969 (Rome, 1970), pp. 3-6.

35. "Celebrating Our Country's Ten Great Years," editorial, Chung-kuo Lin-yeh [Chinese Forestry], Peking, No. 19, 1959, pp. 2-7.

36. JMJP, January 8, 1964, p. 2; and September 26, 1964, p. 2.

37. "Tree Planting in China," Peking Review, No. 17, 1972, pp. 17-18.

38. JMJP, January 23, 1960, p. 1; news release, China News Service, November 16, 1964, p. 4.

39. "Conscientiously Protect the Forests," editorial, JMJP, May 29, 1962, p. 1; Ministry of Forestry, People's Republic of China, "Summing up Forestation Work in the Past Thirteen Years," Chung-kuo Lin-yeh, No. 5, 1963, pp. 1-5.

40. "Regulations Governing the Protection of Forests," released by NCNA, Peking, June 22, 1963; "It Is Everybody's Responsibility to Protect Forests," editorial, JMJP, June 23, 1963, p. 1.

41. NCNA, Peking, May 23, 1965.

42. J. Lossing Buck, Land Utilization in China (Nanking: University of Nanking, 1937), p. 274.

43. Kao Hui-min, "Crop Rotation Systems in China's Agricultural Production," Hung Ch'i, No. 24, 1961, pp. 18-24; Kao Hui-min, "A Discussion on the Coordination Between Agriculture and Animal Husbandry in Relation to the Crop Rotation System," K'o-hsueh T'ung-pao, No. 4, 1964, pp. 283-87.

44. Bureau of Agriculture, ed., I-nien San-shu Ta-yu K'o-wei [Three Crops a Year, A Great Possibility] (Canton: Kwangtung People's Press, 1973).

45. Pai Szu-chiu, "How to Produce Two Abundant Crops in a Two-Crop Rice Culture Program?" Chung-kuo Nung-pao Tseng-k'an [Supplement to Chinese Agricultural Bulletin], Peking, No. 2, 1958, pp. 1-3, 6.

46. Wang Shou, "A Discussion on the Growing of More High-Yielding Crops," in 1956-1967-nien Ch'uan-kuo Nung-yeh Fa-chan Kang-yao (Hsiu-cheng Ts'ao-an) Chiang-hua [Lectures on the National Program for Agricultural Development, 1956-1967 (Revised Draft) (Peking, 1958), pp. 90-94.

47. Information supplied to the author by J. Lossing Buck.

48. JMJP, September 11, 1958, p. 1.

49. Land Utilization Bureau, Ministry of Agriculture, op. cit.

50. Ch'eng Chao-hsuan, "Basic Experience of High-Speed Development of Agricultural Science in China," op. cit.

51. Chu Chien-nung, T'u-ti Fei-li Ching-chi Yuan-li [Economic Principles of Land Fertility] (Shanghai, 1964), pp. 281-91.

52. Research Institute of Mechanical Science, Chinese Academy of Sciences, "The Use of Explosives in Agriculture," K'o-hsueh T'ung-pao, No. 13, 1960, pp. 408-11.

CHAPTER

14

PLANT BREEDING

Because of its large territory and wide variation in natural conditions, China has a great wealth of plants. Many kinds and varieties of plants thrive and are used for food or other purposes. However, little scientific work has been done to evaluate the overall plant resources of the country and to undertake intensive research programs for the improvement of species and varieties. Since the country depends so much upon rice and wheat for food and cotton for clothing, more attention has been given to the improvement of these crops. Rice breeding in particular was a field in which considerable progress had been made prior to the Communist regime.

The NPAD provided that the use of all sorts of existing improved strains of farm crops whose adaptability to local conditions had already been proved by selected experiment should be extended before 1962. In the case of improved strains whose use had in the main been popularized (cotton, for example), steps should be taken to strengthen the work of invigorating and varying strains. Major efforts should be made to cultivate new, improved strains from other parts of the country and from abroad.

The program further provided that the APCs (now communes) should set aside land especially for growing seed, should strengthen the work of seed selection among the masses, and should institute a system for propagating improved strains. In areas with unreliable harvests, attention should be paid to storing up seeds of good strains. State farms should make themselves centers for propagating good strains of farm crops, and should take energetic steps to increase the available supplies and extend the use of improved strains of crops suited to local conditions. All provinces, municipalities, autonomous regions, special administrative regions, autonomous chou, and hsien were urged to set up special seed agencies. (Article 8).

At the outset of the Communist regime, the progress of the seed selection program was rather slow. During the Economic Recovery Period (1949-52) emphasis was given to the selection and popularization of lines developed by the farmers. In the First Five-Year Plan period (1953-57) the exchange of improved seeds between provinces and between hsien was encouraged, and there was popularization of 380 new varieties of rice, wheat, corn, millet, kaoliang, sweet potatoes, soybeans, cotton, and other crops developed by agricultural colleges and research institutions and 104 varieties developed by farmers.

Like other major agricultural activities the seed improvement program was speeded up during the Leap Forward campaign. Agricultural universities, research institutions, and peasants throughout the country selected and cultured more than 1,000 improved varieties. In addition to improved varieties selected from within China, 72 improved varieties had been introduced from abroad by the latter part of 1959, including varieties of cotton, sunflowers (from Hungary), and sugar beets (from the Soviet Union and Poland). The acreage sown to these varieties was expanded steadily.[1]

At the second national conference on the cultivation of crop strains, held at Peking in January 1964, 146 strains of 22 crops developed by Chinese scientists were recommended for propagation. The crops included rice, wheat, corn, sorghum, millet, sweet potatoes, soybeans, cotton, tobacco, sugar beets, sugarcane, rapeseeds, sesame, peanuts, and hemp.[2] Prior to 1949 China used mainly imported strains in the cultivation of cotton, potatoes, tobacco, sugarcane, and sugar beets.

Some 1,600 model stations for the multiplication of improved seeds were reported to have been set up in practically all hsien in 15 provinces and municipalities by the end of 1963. In addition many production teams had established special farms for selecting and preserving improved seeds.[3] By August 1965 more than 1,780 hsien had established stations for multiplying seeds of improved strains, nearly 10,000 communes or production brigades had established improved seed stations or teams, and about one-half of the production teams had set up seed fields.[4] By early 1972 seed selection farms had been set up in practically all hsien and communes, and seed fields and plots had been established in every production brigade or production team. According to incomplete statistics for 22 provinces, municipalities, and autonomous regions, there were more than 7 million technicians from among the masses trained in special schools or courses for seed improvement.[5]

Table 14.1 shows the total acreage and percentage of the area sown to improved seeds of all crops in 1952 and 1955-59. Figures for more recent years are not available, but it is generally believed that

TABLE 14.1

Area Sown to Improved Seeds, 1952, 1955-59

Year	Area Sown to Improved Seeds (thousand mou)	Area Sown to Improved Seeds as a Percentage of Total Sown Area
1952	126,680	6.2
1955	878,000	38.7
1956	1,006,000	42.1
1957	1,200,000	55.9
1958	1,750,000	76.0
1959	1,800,000	80.0

Sources: Liu Ting-an, "Great Achievements in Seed Improvement Work in China in the Past Decade," *Chung-kuo Nung-pao* [Chinese Agricultural Bulletin], Peking, No. 15, 1959, pp. 22-23, 27; T'an Chen-lin, "Strive for the Fulfillment, Ahead of Schedule, of the National Program for Agricultural Development," a report at the Second Session of the Second National People's Congress, April 6, 1960, *JMJP*, April 9, 1960, p. 1; State Statistical Bureau, *Ten Great Years* (English ed.; Peking, 1960), p. 128.

there was an acute shortage of seeds of all kinds during and shortly after the three consecutive years (1959-61) of crop failure and that the situation improved gradually in the early 1960s. In 1965 the acreage sown to improved seeds of rice, wheat, corn, kaoliang (a kind of grain sorghum), millet, and soybeans amounted to one-half of the total acreage sown to these six principal crops and was nearly 200 million mou more than in 1964.[6]

Table 14.2 shows the percentages of acreage sown to improved seeds of various principal crops in 1952-58 and in 1973. In 1952, the last year of the Economic Recovery Period, the area sown to better seeds for cotton was already 50.2 percent. This percentage was far ahead of that for any other major crop because, prior to World War II, improved varieties of cotton had been introduced from abroad, mainly from the United States, and popularized over a large area of China. The percentages of acreage sown to improved seeds for all major crops (rice, wheat, miscellaneous grains, potatoes, cotton, and oil-bearing crops) rose steadily and rapidly during the First Five-Year Plan period (1953-57). The even more conspicuous rise in 1958 was again attributed by the Communists to the Leap Forward campaign. In that year the area sown to better seeds of cotton reached 97.0 percent, and that for other major crops ranged from 61.6 percent for oil-bearing crops to 86.1 percent for wheat.

Claims about the improvement of individual crops in recent years are too numerous to mention; only the three most important crops—rice, wheat, and cotton—are briefly discussed here. Rice was first planted in China some 50 centuries ago. The cultivation was broadened in the twenty-sixth to twenty-second centuries B.C. By the time of the Chou dynasty (1111-256 B.C.), many writings on the cultivation, fertilization, seed selection, irrigation, harvesting, processing, and storage of rice had appeared. Today rice is China's most important crop in production and its acreage is substantially greater than that of wheat.

Many new high-yielding varieties of rice are reported to have been developed since 1949, but some of the varieties developed prior to the Communist regime remain the most widely planted. For example, the cultivation of Nant'ehao (Southern Special), an early-ripening, disease-resistant variety developed in Kiangsi Province before World War II, has been extended to south China and the middle and lower reaches of the Yangtze River. The total acreage of Nant'ehao in these areas in 1961 was estimated at 50-60 million mou and the yield averaged 500-600 chin per mou, with the highest yield exceeding 1,000 chin per mou. In 1974 about 80 percent of all paddy fields in China had adopted improved rice strains.[7]

In the development of new rice strains, special attention has been given to those that ripen in 110 days and are suitable for cultivation in cold northeast China, and to those with short stalks, which withstand typhoons and heavy rains and are suitable for cultivation along the coastal areas.[8]

The cultivation of wheat as one of China's principal crops was recorded as early as the Yin dynasty (1384-1112 B.C.). Today more than one-third of the Chinese population consumes wheat as its most important staple food. More than 100 new varieties are reported to have been developed since 1949, and some of the varieties introduced prior to the Communist take-over have been popularized. Nanta 2419 winter wheat, originally planted only in the Yangtze River basin, now is also sown in the spring wheat areas of the northeast and northwest. This variety is insect-resistant because the inner and outer glumes of the plant's ears are closely woven, so that wheat aphids have no place to lay their eggs.[9] Pima No. 1 winter wheat was grown over only a few small areas in the Yellow River basin in 1950. Its acreage reached 90 million mou in 1959. Compared with other strains, Pima No. 1 has bigger ears, heavier grains, thinner husks, higher flour yield, harder stalks, and stronger resistance to smut and spring cold. In northwest China the yield from Pima No. 1 winter wheat averaged 200-300 chin per mou in 1961, with the highest yields reaching 700 chin per mou. In 1973 wheat of good strains accounted for about 90 percent of the

TABLE 14.2

Percentage of Area Sown to Improved Seeds of Principal Crops, 1952-58, 1973
(total area sown to each crop-100)

Year	Rice	Wheat	Coarse Grains	Potatoes	All Grains	Cotton	Oil-Bearing Crops
1952	5.4	5.1	5.0	0.4	4.7	50.2	1.9
1953	7.9	7.4	8.0	2.2	7.4	61.4	2.4
1954	12.0	23.5	12.9	9.9	14.9	67.7	2.9
1955	19.0	32.7	16.5	13.8	20.6	70.5	4.0
1956	41.3	58.7	21.4	38.3	36.4	89.5	31.5
1957	62.9	68.7	42.5	56.5	55.2	93.9	47.7
1958	81.9	86.1	67.9	81.5	77.5	97.0	61.6
1973	80.0	90.0	n.a.	n.a.	n.a.	n.a.	n.a.

Note: n.a. = not available.

Source: State Statistical Bureau, Ten Great Years (English ed.; Peking, 1960), p. 131; NCNA, Peking, October 29, 1973.

wheat fields in China's 11 wheat-producing provinces and municipalities.[10]

Although China is one of the leading cotton producers in the world, cotton is not a plant native to China. During the T'ang dynasty (618-905) only mulberry and hemp were mentioned in Chinese writings as the country's principal fiber crops. It was not until the end of the T'ang dynasty that cotton was introduced to China from Arabia via Iran and Central Asia to Sinkiang and Kansu. Between the Sung (960-1126) and Yuan (1206-1367) dynasties cotton also was introduced from India over a sea route to Kwangtung, Kwangsi, Fukien, and Yunnan. Later cotton growing spread to central China, and during the Ming dynasty it spread north of the Yangtze River.[11]

Toward the end of the nineteenth century, American cotton was introduced into China by a few missionaries and industrialists. Thereafter the United States played an increasingly important role in developing cotton production in China by providing technical advice and superior seeds. When the Communists took over in 1949, the acreage of American cotton planted in China totaled 10.5 million mou (1.75 million acres), or more than one-fourth of the total cotton area of the country.[12]

Since 1949 cotton planting in China has undergone two major changes. In 1950-55 the planting of Stoneville 4, Deltapine 14, and Coker 100, all of which were introduced from the United States, was further extended to replace native cotton and degenerated foreign cotton. In 1956-60 Deltapine 15 was used in the Yellow River basin and the Yangtze River basin to replace Stoneville 4, Delfos 531, and Acala, while such early-ripening varieties as Chinyü No. 5, 611 P'o, and Kuochi No. 1 were popularized in areas where the growing season is short and certain new long-fiber varieties were introduced in areas where the growing season is long and the temperature is high. In recent years more than 20 new cotton varieties have been popularized in various areas. In the meantime Deltapine 15 had been extensively planted in Kiangsu, Chekiang, Anhwei, Hupeh, and Hunan to replace other Deltapine varieties that had become impure or had degenerated.[13]

In 1956 the Ministry of Agriculture established a network of cotton experiment stations for the study of new varieties developed in China or introduced from abroad. Since 1958 the project has been undertaken by the Cotton Research Institute of the Chinese Academy of Agricultural Science in collaboration with more than 70 agricultural research units and state farms throughout the country. By the end of the first half of 1962, 18 varieties had been selected for popularization in various regions. Thirteen of these were developed in China, while the other five were introduced from abroad. Of these varieties Hsuchou 209 and Shihtuan 5 have done particularly well. In 1961, 100,000 mou of Hsuchou 209 were planted in Honan, Hopei, Shansi, and Shensi, while 360,000 mou of Shihtuan 5 were planted in the Shihchiachuan District of Hopei.[14]

It has been suggested that the PRC may have used the high-yielding strains of rice and wheat developed by the International Rice Research Institute in the Philippines and the International Maize and Wheat Improvement Center in Mexico, which have been used by many countries in increasing quantities. Although both institutions are financed by the United States, seeds of the new strains they have developed are available to all countries. The description of some of the new strains used in the PRC in recent years seems to fit the strains developed by the two institutions.[15] However, in May 1971 leaders of the communes near Peking, Shanghai, and Canton told foreign visitors that they had never heard of the "miracle" rice and wheat developed in the outside world and that for high yields they relied on new strains developed in their own areas.[16]

As already mentioned, plant breeding, especially rice breeding, had made considerable progress in China prior to the Communist regime, and development in this field stood well ahead of that in other lines of agricultural science. Numerous studies in recent years by Ting Ying (1888-1964), a well-known rice breeder, and others indicated that the PRC had continued to make substantial progress in the development of new strains of principal crops. As early as the 1950s, for example, the PRC was already paying much attention to the development of early rice strains that are high-yielding and can withstand typhoons and heavy rains.[17] It is possible that the PRC has developed new strains similar to those developed by the two American institutions. During the visit of the American delegation of plant scientists to China in 1974, Dr. John L. Creech, director of the National Arboretum, Washington, D.C., was told by the Chinese that they had had their own short-strawed rice varieties since the late 1950s. Although the Chinese did bring in IRRI strains during the 1960s, they did not necessarily like them because of limited cold tolerance.[18] Regarding wheat, a Canton publication (1974) revealed that in 1972, more than 30 Mexican varieties of wheat were introduced and popularized in Kwangtung Province with satisfactory results and in 1973, additional Mexican wheat varieties were sown on more than 90,000 mou (15,000 acres) in the province with varying degrees of success. (Mo-hsi-ko Hsiao-mei Tsai-p'ei Chi-shu [Technique of Cultivating Mexican Wheat], edited by the Seed Station, Kwangtung Provincial Agricultural Bureau, and the Food Crops Research Institute, Kwangtung Provincial Academy of Agricultural Science; published by the Kwangtung People's Press, Kwangchou, 1974, the editor's note and pp. 2-3).

But, even if the PRC has succeeded in plant breeding, the mismanagement of seeds has been a serious problem. In 1959 Liu Ting-an, director of the Seed Administration of the Ministry of Agriculture, admitted that the development had been too hasty and that "there was a lack of healthy systems of multiplication, inspection, storage, transportation and utilization, thus creating much confusion."[19]

A number of good strains were reported to have degenerated because of natural mixing with inferior strains, unfavorable climatic conditions, or improper cultivation techniques. Among them are Victory 100 sweet potatoes, Nant'ehao paddy rice, and Deltapine cotton.[20] The seeds in many cotton fields have become impure and have degenerated to varying degrees, thus adversely affecting output and quality. In 1963 Deltapine cotton gave about 34 percent pure cotton as compared with over 38 percent in the previous years. Compared with the past, three to five chin of ginned cotton were lost for every tan of raw cotton, and the quality of cotton was lowered by one or two grades. In Hsinchou chü, Hsinchou hsien, Hupeh, the total output of unginned cotton in 1962 exceeded that in 1957 by 50,000 chin, but ginned cotton was 140,000 chin less than in 1957.[21]

Mismanagement has resulted in an acute shortage of seeds, although this was not openly admitted in a general statement until the beginning of 1961, when the situation became very serious. An editorial in the People's Daily, speaking of the preparation of seed for spring sowing, stated that after two years of natural calamities this was a difficult task. Calamity-stricken areas had no seeds left or had only seeds of very low quality for spring sowing. Moreover, the shortage of seeds was not confined to regions that had suffered natural disasters. In Hopei, Shansi, Kirin, Hupeh, Hunan, Ninghsia, and other provinces, many of the high-quality seeds that had been propagated a few years earlier had been spoiled by heat, insects, or moisture and could not be used for sowing. In some areas high-quality seeds had deteriorated because they were sown with other types of seeds. In other areas the high-quality seeds were not well selected at the time of harvest, or people in charge did not care.[22]

During the Leap Forward campaign party cadres in many localities urged or ordered the communes to discard certain inferior seeds in favor of the improved seeds, even though the latter were not available in sufficient quantities. This zeal to introduce and popularize new varieties of crops, in many cases totally ignoring local conditions, worsened the seed situation and drew much criticism. In an article entitled "Guidance of Agricultural Production Should Be Started from Practicality," Lu Li stated that seeds of inferior varieties should not be discarded at will before better seeds were available:

> We should positively breed new varieties and at the same time continue to use old varieties. For example, in some mountainous areas, there is a type of bean called "kun shan chi" (mountain-rolling pheasant), which is generally

> regarded as an inferior crop. But because it grows on any soil, this crop should not be discarded. Without it, some areas would have no harvest of any kind. Another example is kaoliang, a grain sorghum. Compared with crops of high yield, one may think that kaoliang should be discarded. But kaoliang is indispensable, particularly in the hilly areas. Its stalks may be used as fuel and for building construction, and for making mats, cooking utensil covers, brooms, and many other daily necessities. Without kaoliang, people's life in some areas would be difficult. Still another example is "wan tou" (field peas). Its yield is much lower than that of wheat and some other crops, but it ripens early and is important not only as food but also as a substitute for soap. It should therefore not be disregarded.[23]

In an article in the Kuang-ming Daily, Ts'ai Hsü and Chang Shu-ch'in of the Peking University of Agriculture suggested that in popularizing seeds, attention should not be confined to a few newly developed varieties and that numerous kinds of "farm household seeds" also should be used according to local geographical and climatic conditions and cultivation systems. A newly developed strain may increase production in certain areas but may produce very little in other areas. On the other hand, "farm household seeds," also known as "local seeds," were developed by peasants to meet certain local requirements and therefore should be included at least as part of the seeds used. These seeds are especially important in provinces or regions where a great variety of local conditions exist. In Hunan, for example, 62.7 percent of all seeds used in the province are farm household seeds because of the wide range among the different localities in latitude, elevation, temperature, soil moisture, number of croppings, sowing period of rice (which extends from the middle of March to the end of June), period of transplanting (from the middle of April to the beginning of autumn), period of harvesting (from the middle of July to November), and growth period of rice (from 110 days to 180 days). A similar situation exists in the Kwangsi-Chuang Autonomous Region. In areas subject to natural calamities, farm household seeds are especially important because of their strong resistance to unfavorable environment.[24] The importance of farm household seeds in China was clearly indicated in a list compiled by the Seed Administration of the Ministry of Agriculture in 1958 (including a supplement compiled in 1959), which covered 2,449 superior varieties of 32 crops. Of these, 1,834 varieties, or three-fourths of the total, were local varieties.[25]

On December 20, 1961, the Chinese Society of Crops was founded as a branch of the Chinese Society of Agriculture. Its first meeting was held in Changsha, Hunan, on December 20-28, 1961, to take up the subject of seed development. The discussion centered on the policy pursued by the government during recent years, a policy that now met wholesale condemnation. Participants in the meeting agreed virtually unanimously that future development of new seeds should be carried out in view of the standard of production over wide areas and should be adapted to different regions; in other words, that it is as important to develop seeds for production over a wide area as to follow the practice of producing special high-yielding seeds.[26]

Commenting on the conference, the Peking journal Scientia stated:

> In recent years there was a one-sided emphasis on the selection and breeding of new types of strain which would give high productivity. This had produced good results, but strains so developed, if used under unfavorable circumstances of fertilization and water supply, might produce not more but less than the ordinary strains used locally. Therefore the questions, what policy our country should take in the future in plant breeding, what standards should be followed in developing new strains, and in general what measures should be taken, drew broad attention at the conference.
>
> Experts from all over the country, putting together their experiences, carried out lively discussions, and on the above questions they reached a relatively unanimous view. The majority held that, in order to guarantee production increase in large areas, the direction of our country's plant breeding work in the future should be toward the development of new varieties suitable for the production in large areas, with both higher and lower levels of production taken into consideration. New varieties should be developed to give much higher production than the original varieties over large areas and for many crops, in order to meet the need of different areas.[27]

This approach would mean that consideration should not be confined to a few isolated areas or to some "high-yielding" varieties, and that plant breeding should be dealt with for the country as a whole and adapted to different circumstances, including local farming systems.

PLANT GENETICS

The development of genetics under the Chinese Communist regime should be reviewed here at least briefly, because the subject is closely linked with the question of plant breeding. Before 1949 Chinese geneticists were mostly followers of the school of Gregor J. Mendel and Thomas H. Morgan. When the Communists took over the country, the "bourgeois" scientific schools were expelled and the Russian Michurin-Lysenko school was introduced as the only approach corresponding to materialist philosophy. The China Michurin Society was established at Peking in 1951, with dozens of chapters throughout the country; and a large number of Michurin Youth Farms were organized by the Young Pioneers.

In the spring of 1952, the Ministry of Agriculture invited A. P. Ivanov, a Soviet specialist in Michurin genetics, to serve as an adviser. Accompanied by Ivanov, a joint technical group of the Ministry of Agriculture, the Chinese Academy of Sciences, the North China Institute of Agricultural Science, the Peking Agricultural University, and other institutions made a survey of agricultural conditions in many areas during April-August 1952. In its report the group urged the intensive study and adoption of Michurin's theories in improving agricultural techniques. The survey was followed by a special training class in Michurin's methods of seed selection and multiplication, which lasted from October 21, 1952, to February 20, 1953, and was attended by more than 1,000 technical workers from all parts of the country, including Inner Mongolia and Tibet. Lectures on Michurin's theories and their application were delivered by several Soviet experts.[28]

In an article on China's achievements in Michurin genetics during the first decade of the Communist regime, Liang Cheng-lan gave a rather detailed account of the work done by Chinese scientists in various major aspects of genetics, including the control of individual plant development, vegetative hybridization in animals and plants, research on predetermined cultivation, and research on intravarietal crossing. He also cited several examples of experiments conducted by advanced farmers on the basis on Michurin principles. But he admitted that the research as a whole was not well organized and "lacked depth."[29]

Tse Te-ming, a genetics specialist of the Chinese Academy of Agricultural Science, stated that whereas foreign data exclusively were used in lectures on genetics in China before 1949, since then more and more experimental results with regard to chromosomes and nucleic acids and preliminary results on artificially induced variations have been obtained in China. According to Tsu Te-ming results of studies of ontogenesis, directive breeding, and sexual and asexual hybridization have become even more varied. Since 1958 Chinese scientists have been carrying out a host of experiments in distant hybridization, an important aspect of genetics. Studies in phase development have been made on wheat, rice, millet, corn, soybeans, cotton, and other principal crops. The results have provided data for improving the techniques of cultivation and seed breeding, and have laid a scientific basis for the introduction of crops from distant places. According to Tsu Te-ming the accuracy of the position that organic matter and environmental conditions are intimately connected could be clearly seen from the analysis of phase development of local strains of wheat, rice, and soybeans.[30]

After the death of Stalin in 1953, Trofim D. Lysenko fell from grace; and in April 1956 he was removed from his post as president of the Soviet Academy of Agricultural Sciences. Four months later, on August 10-25, 1956, a conference on genetics was held at Tsingtao under the sponsorship of the Chinese Academy of Sciences and the Ministry of Higher Education. A large number of biologists, animal breeders, plant breeders, and geneticists attended; and some who had maintained silence for years were called upon to address the audience. One of the speakers, Fang Chung-hsi, professor of biology at Shantung University, who was a Michurinist prior to the conference and had written several articles supporting Lysenkoism, revealed that the following basic points were clarified during the two-week discussion:

1. The controversy between Michurinists and Mendelists is not a struggle between materialist and idealist biology, as claimed by Lysenko. The difference between the two schools is a scientific problem, not a philosophical one and still less a political one. Morganists are not reactionary.

2. Morganian genetics was never "beaten down" by Lysenko. In fact, it has developed to a highly refined level far beyond chromosomology. It should not be discriminated against or prevented from further development.

3. There is some justification for the theory of unification of the organism with its living conditions, as claimed by the Michurinian school; and Lysenko's theory of the phase development of plants, although not perfect, is not without foundation. However, these should be further investigated, particularly by refined experimental methods.

4. The recognition of the possibility of directed cultivation does not exclude Mendelism, mutation, or random variation. Proper utilization of existing variations can achieve the purpose of directed cultivation. A dictatorial attitude is harmful to the development of science.

5. Vegetable hybridization and the inheritance of acquired characteristics may be explained in various ways; none of the facts cited by Lysenko invalidates the chromosome and gene theory.

6. Darwin's theory of natural selection is still correct. Lysenko's contention that there is no competition within a species is not adequately supported by evidence.[31]

Although advocating further research and application of Michurin's theories, Tsu Te-ming maintained that the Mendel-Morgan school should not be ignored and that some of its theories definitely should be used for reference.[32] In his opinion the questions in dispute between the Morgan and Michurin schools had changed somewhat in the late 1950s. The two schools had shown a tendency to approach one another, although some differences continued to exist in their viewpoints on the relationships among the material basis for heredity, hereditary variation, and environmental conditions; the hereditary nature of acquired properties; the possibility of directive breeding; and other important problems.[33]

Lysenko somewhat regained his influence in Soviet biological and agricultural circles in 1960 and was again appointed president of the Soviet Academy of Agricultural Sciences in August 1961. He resigned from the presidency on April 5, 1962, for "health reasons."[34] However, since the 1956 Tsingtao conference Chinese Michurinists have been losing ground and Mendelists have increasingly dominated the scene. In the third-year biology course at Peking Normal University, theories of both schools of genetics were taught in the academic year 1960-61.[35] Professor Li Ching-hsiung of Peking Agricultural University, stated: "For some years now, Morgan's theories of genetics have enjoyed the sympathy and patronage of Party and government."[36]

In a review of genetics work in China in 1961 and 1962, Fang Chung-hsi revealed that many essays on both schools were published; and forums were held in Kwangtung, Fukien, Chekiang, Kiangsu, Shantung, and other provinces. He stated:

> Through discussions, we have realized that both Michurinian and Morganian schools have their strong and weak points, and one is not contradictory to the other. We have not only better understood the development and trends of genetics in the world, thus raising the standard of our career and broadening the scope of our knowledge, but also are more aware than before that "Let a hundred flowers bloom and let a hundred schools contend" is our Party's basic policy for developing science and culture, a means to direct research work under Marxism and Leninism, and a mass line of conducting scientific work.[37]

NOTES

1. Liu Ting-an, "Great Achievements in Seed Improvement Work in China in the Past Decade," Chung-kuo Nung-pao [Chinese Agricultural Bulletin], Peking, No. 15, 1959, pp. 22-23, 27.
2. NCNA, Peking, January 27 and 30, 1964.
3. JMJP, December 1, 1963, p. 1.
4. Ibid., August 3, 1965, p. 1.

5. "Selection and Popularization of Good Seed Strains," Peking Review, No. 22, 1972, p. 22.

6. Ibid.

7. NCNA in Chinese, Peking, October 16, 1974.

8. Shen Hsueh-nien, Shui-tao [Paddy Rice] (Peking, 1955), pp. 3-4, 18-28; Tseng Kwang-hui, "Some of China's Superior Strains of Rice," Nung-yeh K'o-hsueh T'ung-hsun [Agricultural Science Bulletin], Peking, No. 2, 1958, pp. 84-86; Cheng Chao-hsien, "Three Years of the Agricultural Charter," China Reconstructs, Peking, No. 10, 1961, pp. 6-10; NCNA, Peking, April 25, 1966.

9. Tai Sung-en, "Fully Develop the Use of Superior Varieties for Increasing Crop Production," Hung Ch'i [Red Flag], Peking, No. 2, 1963, pp. 25-34.

10. NCNA-English, Peking, October 29, 1973. For a brief history of China's wheat production and a description of some of the wheat varieties popularized in China in recent years, see Laiyang Agricultural School, Laiyang, Shantung Province, Hsiao Mai [Wheat] (Peking: Science Press, 1975), pp. 2-7, 219-26.

11. Hu Ching-liang, Chung-kuo Mien-ch'an Kai-chin-shih [History of the Improvement of China's Cotton Production] (rev. ed.; Shanghai, 1947).

12. Feng Tse-fang, Ho-yu Chung-kuo Tsai-chung ti Hsi-jung-mien [Fine Velvet Cotton Suitable for Plantation in China] (Shanghai, 1953), pp. 8-9; Mu Chen, "Cotton-Producing Areas in China," Ta Kung Pao, Peking, April 5, 1962, p. 2.

13. Ch'en Jen, "Multiplication and Popularization of Superior Cotton Varieties," JMJP, May 28, 1963, p. 5.

14. Jen-min Shou-ts'e [People's Handbook] (Peking, 1962), p. 274.

15. New York Times, October 26, 1969, p. 21; The Green Revolution, Proceedings Before the Subcommittee on the National Security Policy and Scientific Developments of the Committee on Foreign Affairs, U.S. House of Representatives, Ninety-First Congress, First Session (Washington, D.C., December 5, 1969), p. 37.

16. New York Times, May 16, 1971, p. 17. Also see National Academy of Sciences, Plant Studies in the People's Republic of China (Washington, D.C., 1975), "The Introduction and Widespread Use of Improved Crop Varieties," pp. 25-26.

17. Shen Hsueh-nien, op. cit.; Tseng Kwang-hui, op. cit.

18. Information supplied to the author by Dr. Creech.

19. Liu Ting-an, op. cit.

20. Sun Chia-tseng, "Let Good Strains Permanently Preserve Their Youth," K'o-hsueh Ta-chung [Popular Science], Peking, No. 4, 1964, pp. 121-22.

21. "Proposals on Several Important Technical Measures for Increasing Cotton Output Next Year," JMJP, August 26, 1963, p. 2.

22. JMJP, November 11, 1961, p. 2; November 25, 1961, p. 1; December 3, 1961, p. 2; January 27, 1962, p. 1; March 15, 1962, p. 2; June 3, 1962, p. 1.

23. Ibid., November 12, 1960, p. 7.

24. Ts'ai Hsu and Chang Shu-ch'in, "Functions of the Farm Household Seeds in the Current Agricultural Production," Kuang-ming Jih-pao [Kuang-ming Daily], Peking, July 30, 1962, p. 1.

25. P'u Mu-hua, "Resources of Crop Varieties of Our Country," JMJP, December 11, 1962, p. 5.

26. Kuang-ming Jih-pao, January 3, 1962, p. 1.

27. K'o-hsueh T'ung-pao [Scientia], Peking, No. 1, 1962, p. 1.

28. "Training Class in Michurin's Methods of Seed Selection and Multiplication," in Ministry of Agriculture, Nung-yeh K'o-hsueh Chuan-t'i Pao-kao chi Ts'an-k'ao Tzu-liao Chi [A Collection of Reports and Reference Materials on the Special Problems of Agricultural Science] (Peking, 1953), pp. 268-301.

29. Liang Cheng-lan, "China's Achievements in Michurin Genetics During the Past Decade," Sheng-wu-hsueh T'ung-pao [Biological Bulletin], Peking, No. 10, 1959, pp. 462-67.

30. Kuang-ming Jih-pao, June 14, 1961, p. 1.

31. C. C. Li, "Genetics and Animal and Plant Breeding," Sciences in Communist China, Publication No. 68 of the American Association for the Advancement of Science (Washington, D.C., 1961), pp. 297-321.

32. Tsu Te-ming, "Development and Accomplishments of Michurin's Genetics in China in the Past Ten Years," Nung-yeh Hsueh-pao [Acta Agriculturae Sinica], Peking, Vol. X, No. 5 (1959), pp. 330-43.

33. Kuang-ming Jih-pao, June 14, 1961, p. 1.

34. New York Times, April 6, 1962, p. 12.

35. Kuang-ming Jih-pao, April 8, 1961.

36. Ibid., May 5, 1961.

37. Fang Chung-hsi, "Positive Influence of Discussions on Genetics During the Past Year or So," Kuang-ming Jih-pao, May 3, 1962, p. 2.

CHAPTER

15

CLOSE PLANTING

The idea of planting crops at a proper density in order to obtain maximum yields is not new in China. In a book entitled Lu-shih Ch'un-ch'iu (Lu's Annals), compiled by Lu Pu-wei, a prime minister in the Ch'in dynasty (255-207 B.C.), the technique of close planting and its requirements and limitations were thoroughly discussed. For centuries Chinese farmers have used such factors as the kinds of crops, land fertility, and local conditions for determining the density of seedlings that would result in higher yields. In general close planting was adopted in China to only a limited extent before the Communist take-over.[1]

At the advice of F. C. Lutsenko, a Soviet agronomist, close planting was encouraged at the beginning of the Communist regime. An experiment was conducted by the Ministry of Agriculture in 1951 at the Shuangch'iao State Farm in the municipality of Peking, where more than 4,000 cotton seedlings were planted on each mou. The yield was reported to have doubled that of the preceding year, although no actual amount was revealed. A similar experiment for wheat was conducted in 1951 at the same state farm, where plots sown with rows 6.5 inches apart produced 80 percent more wheat than those sown 18 inches apart. Cadres in charge of agriculture in various provinces were later sent to Peking to study the technique of close planting.[2] At the same time close planting of rice was practiced in the Po Hai area of north China, with a reported yield of more than 1,000 chin per mou, as compared with the average yield of rice in the PRC in 1951, which was 300 chin per mou.[3]

Despite the reported success of these experiments, close planting was opposed by many scientists attending the January 1951 North China Conference on Agricultural Techniques. As a compromise the conference recommended increasing the amount of seed sown in each row but widening the distance between the rows. Many farmers also doubted the wisdom of close planting, mainly because they could not afford the seeding and transplanting machines used by the state farms.

It was claimed that in 1955 some 150 million mou of wheat farms, or 37 percent of the land sown to wheat, adopted close planting, and that in most of the remaining wheat-growing areas traditional densities were modified to bring them nearer to close planting. The increase of the country's wheat production from 18.1 million metric tons in 1952 to 24.8 million metric tons in 1956 was partly attributed to the wide adoption of close planting throughout the country.[4]

Actually, however, during the First Five-Year Plan period close planting was continued with only moderate enthusiasm. It was during the Leap Forward campaign that this practice of cultivation was again emphasized. In 1958 from 20 to 30 chin of wheat seeds were sown to one mou in many areas, and the resulting harvest was reported to be from 400 to 600 chin per mou, compared with yields of 200 to 300 chin per mou at the normal density of planting, 6 to 15 chin per mou. For rice the target was from 35,000 to 40,000 clusters and from 350,000 to 400,000 seedlings per mou.[5]

Apparently because of the failure to increase agricultural production by the indiscriminate practice of close planting, a series of articles on the problem was published in the leading newspapers and scientific journals in 1959-61. Although some writers still placed an emphasis on close planting as a means of raising yields of the principal crops, many scientists warned that it should be practiced only to a certain degree, according to particular circumstances.

In the case of wheat, a survey of 183 experimental plots of high-yielding winter wheat in the municipality of Peking and the provinces of Hopei and Honan, conducted in 1958 by the Research Institute of Plant Breeding and Cultivation, indicated that close planting beyond a certain limit had resulted in a decrease in yield because of a serious interference with normal growth of the plants. It was suggested that close planting should be practiced according to the conditions of planting, fertilization, irrigation, light, temperature, moisture, seed variety, and other factors, and that the proper degree of close planting varies to a certain extent because of differences in these factors.[6]

Another experiment conducted in 1958 by the Kiangsu Provincial Branch of the Chinese Academy of Agricultural Science in Nanking indicated that at the density of more than 500,000 wheat seedlings per mou, the normal development of both individual and plant community as a whole was upset, mainly because of progressive shading of the leaves. Under such circumstances the fundamental basis for higher yield was lost. It was suggested that a density ranging from 250,000 to 350,000 seedlings per mou would seem to be more practicable.[7]

A similar conclusion was drawn from an experiment on wheat conducted in 1959-60 by a special team of the Research Institute of Phytophysiology of the Chinese Academy of Sciences in collaboration with the Agricultural Research Institute of the Hsinhsiang Special District in Honan Province. The results indicated that after the density of planting had been increased to a certain point, the "economic coefficient" of yield would tend to decline.[8]

In the case of rice, experience with close planting has been essentially the same as that with wheat. According to a survey of 335 farms in Kiangsu Province conducted in 1958, there were some cases where yield was higher than normal when 400,000 to 600,000 seedlings were planted to one mou. But when the number of seedlings per mou exceeded 600,000 the average number of grains in each ear tended to decrease.[9]

In a series of 11 experiments on the density of planting of corn conducted in northern Kiangsu in 1958, the following results were reported by the participants:

1. The optimum density of planting varied with "agrotechnic levels." Under the conventional agrotechnic level, higher yield for the variety with a stalk of 200 centimeters or more in height could be reaped by a density of 3,000 plants per mou in spring sowing. Under a higher agrotechnic level it was preferable to run as many as 4,000 plants per mou in both spring and summer sowing to obtain the best results. It was indicated that within the range between 2,000 and 8,000 plants per mou, the closer the density, the higher the yield per stalk.

2. There was a close relationship between the grain yield per unit area of land and agrotechnic level. The average grain yield per mou treated under higher agrotechnic level exceeded that under conventional agrotechnic level by 50 to 63 percent. Under higher agrotechnic level the reduction of grain yield as a result of too close spacing was more slack than that under conventional agrotechnic level.

3. The addition of manure at a heavier density in proportion to the planting also gave a slacker reduction than planting spaced too closely without a corresponding increase of manure.

4. The heavier density of planting increased the total number of ears per mou, but the rate of increase was not as great as that of the increased plants. For instance, if the number of plants per mou increased by four times, from 2,000 to 8,000, the number of ears per mou increased by only 2.26 to 2.62 times. Furthermore, under too closely spaced conditions the number of kernels per ear and the weight per 1,000 grains were significantly decreased and a lower yield per unit area of land was obtained.

5. By increasing the density of planting, it was apparent that the height of plant and ear was neither increased nor decreased but the stem became markedly slender, thus resulting in a higher percentage

of stalk breaking. The increased planting density also tended to delay tassel emergence and ear silking. For instance, in the case of 8,000 plants per mou, tassel emergence and ear silking were delayed by two days and three to six days, respectively, as compared with the case of 2,000 plants per mou. With regard to root development, with increased density of planting the number of permanent roots of a single plant was reduced and the distribution was shallower. In the case of 8,000 plants per mou, the depth of permanent roots and the distribution of a single plant's roots were reduced by 36.18 percent and 34 percent, respectively, as compared with the case of 2,000 plants per mou. Furthermore, under closely spaced conditions the accumulation of plant food materials was reduced and the development of underground parts was thereby inhibited. This inhibition, in turn, affected the accumulation of organic matter.

6. The differences of temperature and relative humidity between plants with different planting densities were remarkable only under such conditions as the entire shading of leaves between rows and great difference of temperature between day and night. With increased planting density relative humidity increased and light intensity decreased.

7. The closer the stands were, the greater the absolute value of the green leaf area. Consequently the power of photosynthesis was increased. However, the process of net assimilation was checked as the green leaf area and the power of photosynthesis increased. For this reason the biological yield was not always raised by increasing the density of planting. Furthermore, under too closely spaced conditions the green leaf area would seriously decrease as soon as its maximum was reached, because of the rapid increase in the lower part of yellow leaves that could hardly stay constant for a certain period.[10]

Among the most heated debates on the question of close planting was the one between Yin Hung-chang, deputy director of the Research Institute of Plant Physiology of the Chinese Academy of Sciences, and Wu Cho-nien, assistant head of the department of agricultural science of the South China College of Agriculture. Yin conducted an experiment in Sunkiang in 1959 with single-cropping late rice of the lao lai ch'ing strain, transplanted on plots that previously were used for growing hung hua ts'ao (red-flower grass). Yin conducted another experiment in Sunkiang in 1959 with the same variety of rice but on a plot where barley had grown previously (see Table 15.1).

The first experiment indicated that despite the great difference in the number of seedlings transplanted, the ultimate effective ears on different plots were "quite near to one another" in number. When the density (number of seedlings per mou) was increased 12 times, the number of effective ears showed less than a onefold increase and the number of grains per ear decreased by almost one-half. The second experiment showed that after the number of ears in a mou had

TABLE 15.1

Results of Yin Hung-chang's Experiments with Transplants of Lao Lai Ch'ing Rice

Transplants of Lao Lai Ch'ing on Plots Previously Used for Red-Flower Grass			
Plot	Number of Basic Seedlings per Mou	Number of Effective Ears per Mou	Number of Grains per Ear
1	30,000	194,000	99.0
2	90,000	243,000	84.4
3	150,000	246,000	74.4
4	210,000	248,000	71.1
5	270,000	266,000	66.7
6	360,000	342,000	56.1

Transplants of Lao Lai Ch'ing on Plots Previously Used for Barley						
			Number of Basic Seedlings (tens of thousands per mou)			
			30	35	40	50
Number of effective ears (tens of thousands per mou)	19.1	21.8	27.0	29.6	31.0	39.0
Number of grains per ear	73.2	69.8	58.6	56.6	52.5	43.5
Percentage of empty grains	12.0	17.0	15.8	23.1	21.4	25.9
Weight of 1,000 grains (grams)	27.6	27.6	27.0	26.6	26.7	26.2
Total number of grains (tens of thousands per mou)	1,398.0	1,522.0	1,582.0	1,675.0	1,628.0	1,697.0
Number of full grains (tens of thousands per mou)	1,230.0	1,263.0	1,332.0	1,288.0	1,280.0	1,257.0
Computed yield (chin per mou)	679.0	697.0	719.0	685.0	683.0	659.0
Verified yield (chin per mou)	829.0	831.0	798.0	784.0	779.0	719.0

increased to a certain extent, the number of grains per ear decreased, the percentage of empty grains increased, and the weight of each 1,000 grains fell off. Within a certain limit the increase in the number of ears exceeded the decrease in the number of grains and in the weight of grains, and therefore the yield still rose along with the increase in density. Having reached that limit, the increase in the number of ears was offset by the decrease in the number of grains and the weight of grains, and therefore the yield no longer increased. After the density had exceeded this limit, the increase in the number of ears was not enough to offset the decrease in the number of grains and the weight of grains, and as a result the yield fell off.

In Yin Hung-chang's opinion a plant group is a system quite capable of self-adjustment. In the plots with thin planting, there are more tillers and the density increases by itself. In the plots with close planting, there are fewer tillers, that is, a natural thinning out is brought about by the death of some of the tillers. Eventually these two kinds of plots are "near to each other" as regards density. Therefore, relying on close planting as a major means of increasing agricultural production is improper: rational close planting can obtain the highest yield at the minimum cost under given conditions of soil, climate, fertilization, irrigation, and labor force.[11]

Wu Cho-nien disagreed with Yin Hung-chang about the latter's theory that a plant group is a system quite capable of self-adjustment and that thin planting and close planting will eventually result in approximately the same density through the increase and decrease of tillering. Using Yin's data (see Table 15.1), Wu maintained that in the first experiment, although the increase in the number of ears from 194,000 to 342,000 per mou did not keep pace with the increase in the number of seedlings from 30,000 to 360,000 per mou "because of cultivation technique," nevertheless the increase of the ears was caused by the increase of the seedlings and was anything but a gift given by the self-adjustment of the plants; Wu stated that close planting definitely had stepped up the yield.[12]

In the 1960s the word "rational" was emphasized in discussing the problems of close planting. In claiming accomplishments in this field, Vice-Premier T'an Chen-lin stated: "Rational close planting has been widely popularized. Too close planting and too sparse planting are both on the wane, and the laws of rational close planting are being grasped by more and more people."[13]

Apparently there has been hesitation about close planting, which was generally promoted during the Leap Forward campaign with some disastrous consequences. "There are people who do not show interest and do not welcome it," said the Kuang-ming Daily.[14] But in principle close planting is still being encouraged. The People's Daily said, "Close planting is an important component of the Eight-Character

Agricultural Charter, but we must remember that 'rational close planting' is what Chairman Mao has said. If we act accordingly, we can increase output and support socialist construction and world revolution."[15]

NOTES

1. Wan Kuo-ting, "The Theory and Methods of Equal Spacing, Close Planting and Maintenance of Seedlings in China 2,200 Years Ago," Nung-yeh Hsueh-pao [Acta Agriculturae Sinica], Peking, Vol. VII, No. 3 (1956), pp. 269-74.

2. Liu Chung-hsuan, "Close Planting Catches on," China Reconstructs, Peking, Vol. V, No. 10 (1956), pp. 22-24.

3. Jen-min Shou-ts'e [People's Handbook] (Peking, 1952), p. 354; State Statistical Bureau, Ten Great Years (English ed.; Peking, 1960), p. 121.

4. Liu Chung-hsuan, op. cit.

5. "What Are the Key Factors in the Realization of a Continued Leap Forward in Agriculture This Year?" Ch'iu Shih [Seeking the Facts], Hangchou, No. 3, March 15, 1960, pp. 1-13.

6. Research Institute of Plant Breeding and Cultivation, Chinese Academy of Agricultural Science, "The Extent of Rational Close Planting of Winter Wheat as Judged from Seedling Growth," Nung-yeh Hsueh-pao, Vol.X, No. 2 (1959), pp. 86-101.

7. Kiangsu Branch of the Chinese Academy of Agricultural Science, Nanking, "Studies on the Rational Sowing Density of Wheat Plants," Nung-yeh Hsueh-pao, Vol. X, No. 5 (1959), pp. 431-46.

8. Yu Shu-wen and Chou Chia-huai, "The Phenomena of Group Concept Development of Wheat Under Different Conditions of Close Planting," in Shang-hai-shih K'o-hsueh Chi-shu Lun-wen Hsuan-chi: Nung-yeh [Collection of Selected Scientific and Technological Essays of the Municipality of Shanghai: Agriculture] (Shanghai, 1960), pp. 35-49.

9. Research Institute of Phytophysiology and Peking Research Office of Phytophysiology, Chinese Academy of Sciences, "Functions of Phytophysiology in Summing up the Experience in the Production Increase of Rice," K'o-hsueh T'ung-pao [Scientia], Peking, No. 6, 1959, pp. 177-78.

10. Fan Fu-jen, Mo Hui-tung, and Ch'in T'ai-ch'en, "A Study on the Close Planting of Corn," Nung-yeh Hsueh-pao, Vol XI, No. 2 (May 1960), pp. 109-36.

11. Yin Hung-chang, "Group Concept in Agricultural Production," JMJP, June 13, 1960, p. 7; and "More on Group Concept in Agricultural Production and Reply to Comrade Wu Cho-nien," JMJP, June 3, 1961, p. 7.

12. Wu Cho-nien, "A Discussion About the Article, 'Group Concept in Agricultural Production,'" JMJP, March 1, 1961, p. 7.

13. T'an Chen-lin, "Stive for the Fulfillment, Ahead of Schedule, of the National Program for Agricultural Development," JMJP, April 7, 1960, p. 1.

14. Ch'eng I-te, "Respect Experience and Stick to Traditions," Kuang-ming Jih-pao [Kuang-ming Daily], Peking, February 26, 1965, p. 4.

15. Joint Investigation Group of the Loyang Area Revolutionary Committee and Military Subregional Command, and Yenshih Hsien Revolutionary Committee and Militia Department," A Lesson on Close Planting, in Which Proletarian Politics Is Brought to the Fore," JMJP, March 28, 1969, p. 3.

CHAPTER 16
PLANT PROTECTION

Losses of agricultural products as a result of pests, diseases, and weeds are difficult to measure. In the United States, a country where plant protection is highly developed, in 1951-60 insects were estimated to cause a loss of 4 to 12 percent in the potential yield of the main grains, diseases a loss of 7 to 14 percent, and weeds a loss of 10 to 17 percent.[1]

Although detailed data are not available on the quantities of crops damaged by insects and diseases in the PRC, it is evident that the losses have been considerable. J. Lossing Buck has estimated that insects and diseases have damaged as much as 10 to 20 percent of grain crops.[2] In 1962, in the areas seriously affected, the loss of grains was about 10 percent of the total production; that of cotton, 20 percent; and that of fruits, 30 percent, according to an estimate made by the Bureau of Plant Protection of the Ministry of Agriculture.[3]

The seriousness of the damage was also indicated in a plan mapped out by the same bureau in 1958 for the thorough control of plant diseases and insect pests on a total of 182.3 million mou (30.38 million acres) of crop land in 18 provinces and municipalities; the area involved included 55 million mou of land for rice, 24 million mou for wheat, 63 million mou for miscellaneous grains (including underground insects), 37 million mou for cotton, and 3.3 million mou for fruits and other economic crops. The Bureau of Plant Protection estimated that the plan, if implemented, would save some 2.5 billion chin (1.25 million metric tons) of grains and 1.8 million tan (90,000 metric tons) of cotton.[4] However, the results of the implementation of the plan were not revealed

The NPAD called for the general elimination, in the 12 years beginning in 1956, of the insect pests and plant diseases that are most harmful to crops, including locusts, rice borers, army worms, corn borers, cotton aphids, red spiders, pink bollworms, wheat juice-sucking worms (Sitodiplosis mosellana Gehin), wheat smut, wheat nematodes, and black rot on sweet potatoes. It also called for local plans for plant protection to include programs for the eradication of other serious insect pests and plant diseases, the strengthening of quarantine systems, the production and improvement of pesticides and related apparatus, and the assurance of their safe and effective use (Article 15).

Also included in the program was the elimination of "four evils" (rats, sparrows, flies, and mosquitoes) in the 12 years beginning in 1956, for reasons of public health as well as agricultural production. When the program was revised in 1960, sparrows were replaced by bedbugs, in recognition of the sparrows' usefulness in destroying insects (Article 27.

In his report to the Second Session of the Second National People's Congress on April 6, 1960, Vice-Premier T'an Chen-lin claimed that the damage caused by aphids and red spiders had been eliminated in the principal cotton-growing areas and that virtually no withered or red leaves could be seen on the cotton plants. Locusts and army worms were basically under control, and damage caused by wheat smut and rice borers had been generally reduced to less than 2 percent. According to T'an Chen-lin, "in many places, blocks of tens and hundreds of thousands of mou of farmland have been freed from diseases and insect pests."[5]

In another account Vice-Minister of Agriculture Yang Hsien-tung stated that in 1962 the total area under control with pesticides was close to 400 million mou (67 million acres). According to Yang Hsien-tung in ten provinces including Honan, Hopei, Shantung, Hupeh, and Kiangsu, more than 4,000 million chin (2.67 million tons) of grain and more than 1 million chin (666.50 tons) of cotton were additionally harvested because of the control of plant diseases and insect pests.[6]

INSECT PESTS AND CROP DISEASES

Following are the highlights of the accomplishments in the control of principal insect pests and crop diseases since 1949, according to Communist accounts.

Locusts

For centuries locusts have been a great menace to China's agricultural production. The earliest outbreak recorded in Chinese history was in 707 B.C., and since then more than 800 outbreaks have taken place. In the past 1,000 years a major locust pest usually broke out once every two or three years, or twice every five years. The frequency is believed to be closely related to the floods once every three years and droughts once every two years on the great plains along the lower reaches of the Yellow River, the Huai River, and the Hai River. The outbreak of East Asian locusts (Locusta migratoria manilensis Meyen) has been concentrated in the coastal areas, the lakeside areas, the inland waterlogged areas, and the flooded areas along the three rivers.

During the Economic Recovery Period (1949-52) more than 120 million persons, mainly peasants, participated in several locust control campaigns. When a major outbreak took place in 1951, the methods of surrounding, whipping, and burning were generally used in one of the most difficult fights against locusts, a fight that was successful despite an acute shortage of insecticides and equipment.[7] In 1951 aircraft were used for the first time in China to apply pesticides for the control of locusts in some areas.[8]

Since 1952 locust control stations have been set up in many areas to conduct reconnaissance work in finding the locust eggs, unfledged locusts, and full-grown locusts. Meanwhile, soil, climatic, and vegetation conditions in many localities frequently affected by locusts have been improved by building water conservation projects and converting the land into paddy fields, salt fields, or afforestation sites.[9] By the use of insecticides and the transformation of breeding grounds, the area affected by locusts had been curtailed by more than 60 percent by the end of 1962, according to Chin Shan-pao, vice-president of the Chinese Academy of Agricultural Science.[10]

However, the menace of locusts to crops remains serious. As admitted in 1963 by Vice-Minister of Agriculture Yang Hsien-tung, although the density of locusts had been reduced, the area over which locust infestation occurred had not yet decreased. The waste in the control work was immense, largely because in recent years many areas weakened their locust control organizations and therefore the facts concerning locusts were not mastered. At the same time many areas had not properly mobilized the masses for locust control. In addition, too much reliance was placed on airplanes for spraying insecticides and airplanes were used for control even where they should not be used.[11] In 1963 about 70 percent of the dusting of insecticides for locust control was done by airplanes.[12]

Rice Borers

Rice borers in China are of two principal types: Schoenobius incertellus Walker and Chilo simplex Butler. Both are generally distributed along and south of the Yangtze River. Two or three generations are produced each year in the Yangtze basin, from three to five generations south of the river, and as many as seven generations on Hainan Island. The degree of their damage to rice crops depends largely on climatic conditions, planting systems, and the varieties of the rice plants used in a particular area. As a result of the improvement of planting systems in the rice-producing regions and the increasing application of chemical insecticides, the loss caused by rice borers in China was reported to have been brought down to less than 2 percent by 1958. Before 1954 tobacco was the principal material used for the control of rice-stem borers. Later the use of "666" was promoted gradually, with good results. In recent years experiments have proved that "1605" is even more effective for the control of borers. Chloropicrin and methyl bromide are used successfully to fumigate the insect pests in warehouses.[13]

Rice borers used to destroy one-tenth of China's annual rice harvest before 1949. As a result of mass cooperation in observing and forecasting the movement pattern of the insect and of control measures taken, damage in 1973 was reported to have been reduced to about 1 percent of the harvest in China's principal rice-growing areas.[14]

However, rice borers apparently remain a menace to rice production. While claiming that peasants of Kiangsu, Chekiang, Anhwei, Hupeh, and other rice-producing provinces had successfully wiped out rice borers on millions of mou of paddy fields, a People's Daily editorial stated that one-sided emphasis was being given to control instead of prevention, to chemical instead of agricultural techniques, and to crash operations instead of consistent measures. According to the editorial, this approach produced only a temporary effect, and the rice borers often multiplied again after being partly eliminated. The result was fluctuation in rice output. The paper urged the communes to prevent as well as control rice borers, with emphasis on prevention; to adopt both foreign and indigenous methods; to combine the efforts of the leadership, the specialists, and the masses; and to carry out measures against rice borers in an overall manner so as to insure maximum efficiency and safety.[15]

Rice Blight

Rice blight (Piricularia oryzae Bri. et Cavara) has affected practically all rice-growing areas in China, particularly those in the north. Three major measures have been taken to control this disease: (1) popularization of disease-resistant varieties of rice, such as "853" and "412," which have been generally used in the T'ai Lake area; (2) proper use of fertilizers, especially nitrogenous fertilizers; and (3) application of pesticides, particularly Ceresan, Mersolite, and slaked lime. In addition, studies have been made on the virus causing the disease, the ecology of rice, the occurrence of leaf blight, and the influence of climatic conditions.[16]

Wheat Aphids

These wheat juice-sucking worms (Sitodiplosis mosellana Gehin) are generally distributed on the low-lying and wet plains along the Wei River in Shensi, the Yi and Lo rivers in Honan, the Huai River in Anhwei, the Chialing River in northern Szechuan, the Yangtze River, and the Han River. Generally there is one generation each year but, because the hatching of the unfledged worms depends on the temperature and humidity of the soil, the dormant period may last as long as several years if the environmental conditions are not favorable for the aphids' growth. It is also known that the larvae leave the soil when the wheat plants draw joints and that the adult insects are in full flush when the wheat plants draw ears. In recent years "666" and other insecticides have been applied on the wheat fields as a means to retard the growth of the unfledged insects in the soil. This measure is reported to have produced remarkable results. Insecticides also are applied to the plants when the adult insects are in full flush. In many areas such insect-resistant varieties of wheat as Nanta 2419 and Hsinung 6028 have been selected for cultivation. The inner and outer glumes of the ears of these two varieties are closely woven, and therefore the wheat aphids have no place to lay their eggs.[17]

Wheat Rust

Wheat rust is among the most serious plant diseases in China, and there has been an outbreak of it virtually every year. In past years when there was a major outbreak of the disease, wheat yield in north China and northeast China often was cut by 30 to 40 percent. It has gradually been ascertained that the outbreak of rust on wheat leaves occurs principally on the plateaus of Kansu and Chinghai. The wheat rust virus thrives on young wheat plants in this area during the summer and in autumn is spread by air currents to the areas along the Wei River and to north China, where it attacks the autumn crops. In spring the disease first breaks out in the Kuanchung area in Shensi, the southern part of Shansi, and the southeastern part of Honan, and then spreads to other areas. During the winter the virus that causes the outbreak of rust on wheat stalks thrives principally on the young wheat plants in Fukien, Kwangtung, Yunnan, and other provinces in south China and later spreads to Kiangsu, Anhwei, and as far as north China and northeast China.[18] More than ten rust-resistant varieties of wheat have been selected from the spring wheat regions in northeast China and northwest China for popularization in other areas. Native agricultural drugs as well as chemical pesticides have been used to control the disease with considerable success.[19] Since the mid-1960s more than 100 rust-resistant strains have been bred and widely adopted. They have proved effective in combating stripe rust. One winter wheat strain bred by the Northwest Agricultural Institute effectively resists diseases. It has been sown on more than 10 million mou in 11 north Ching provinces and has yielded roughly 20 percent more than ordinary strains.[20]

Cotton Pink Bollworms

With the exception of the Sinkiang-Uighur Autonomous Region, western Kansu, and certain hsien in the northern parts of Shansi, all cotton-producing areas in China have been invaded by pink bollworms (Pectinophora gossypiella Saunders). The number of generations each year varies among different regions from two to five, depending on climatic conditions. A notice issued jointly by the Ministry of Agriculture and the Ministry of Commerce on September 19, 1960, stated that since the Leap Forward campaign, damage caused by bollworms had been basically eliminated in most cotton-growing areas in northeast China and areas north of the Yellow River, and that damage suffered by cotton-growing areas in the Yangtze valley had been greatly alleviated. However, the notice added that on the whole the work of eradication was "still not balanced well enough."

The ministries urged the cotton-growing communes and state farms to pay special attention to the eradication of bollworms in cotton that is being dried or temporarily stored. Communes and state farms that use their own seeds were asked to have the seeds well smoked to eradicate the pests. The ministries also directed that, where feasible, the method of eradicating bollworms by natural freezing, which had been popularized in northeast China, be tested in other areas. Various localities were urged to adopt the pest-preventing types of warehouses used in Hupeh, Kiangsi, and other provinces, to use lime of cement to build protective walls, and to spray "666" powder on the walls.[21]

Cotton Aphids

The degree of damage caused by cotton aphids (Aphis gossypii Glover) in China varies among regions according to climatic conditions, and such damage is usually heavier in the north than south of the Yangtze River. In the early years of the Communist regime, tobacco water, cottonseed oil soap, and other vegetable insecticides were used for the control of cotton aphids. Their effectiveness was low and not durable. In 1952 the use of 0.5 percent "666" powder, "1605," and "E-1059" was introduced. These chemical insecticides were reported to have produced good results in controlling other cotton insects as well as aphids.[22]

Late Potato Blight

The outbreak of this disease (Phytophthora infestans [Mont.] De Bary) in China was very serious in 1950, 1952, 1953, and 1954 because of the heavy precipitation and high humidity in the potato-growing areas. In north China, northeast China, and northwest China, where potatoes generally are sown in the spring and harvested in the autumn, serious damage by this disease often occurs after the heavy precipitation in June and July. In east China and southwest China, where there are two crops of potatoes each year, the first crop often suffers heavy damage from the blight as a result of the high humidity in the spring.[23] As part of a study on the potato-growing areas in northern Shansi, conducted jointly by several agricultural research institutions in 1958, a method of controlling the late blight by eradicating the central diseased plants and applying copper sulfate solution, Ceresan, lime, and other insecticides was experimented with successfully.[24]

Army Worms

Army worms (Cirphis unipuncta Haworth) may be found in virtually all parts of China. They travel in great armies, damaging grains, grasses, and other crops. In northeast China there are two or three generations each year; in north China, three or four generations; and in east China, as many as eight generations. Reportedly the use of DDT, "666," and other insecticides for killing the worms has been successful.[25]

PESTICIDES AND EQUIPMENT

Before the Communist take-over, China had only three factories manufacturing chemical pesticides. These factories were located in Shanghai, Peiping, and Shenyang and were operated by the Ministry of Agriculture and Forestry of the Nationalist government.[26] Since 1949 the number of pesticide plants has increased steadily. Among the large newly constructed plants are those in Shanghai, Tientsin, and Changtien (Shantung Province).[27]

Copper sulfate was the only chemical pesticide produced in China prior to World War II; and its total production in 1949 was 64 tons, according to a Communist account. It was claimed that in 1957, the last year of the First Five-Year Plan, the output of chemical pesticides reached 64,700 tons and that in 1958, 65,000 tons of "666" (benzene hexachloride, or BHC) and 4,700 tons of DDT were produced. The first pesticide produced on a large scale was "666," and its production increased about 110 times between 1952 and 1958. Trichlorfon (Dipterex) was produced on an experimental basis in 1957. Other pesticides produced domestically in considerable amounts in 1957 included Demeton (E-1059), "1605," and chlordane.[28]

In 1973 China had about 300 factories producing some 120 kinds of chemical pesticides (insecticides, fungicides, and herbicides), as against 20 kinds in 1957, 30 kinds in 1959, and 60 kinds in 1963. Most of China's provinces, municipalities, and autonomous regions now have small or medium-size pesticide factories. There are also more than 20 research institutes in this field with a total of 1,000 research workers. Special attention has been given to developing pesticides of low toxicity and those that can be made from industrial by-products. Production of organic phosphorus compounds also has been emphasized because these are highly effective and can be used in various ways against many kinds of pests.[29]

Because of the increasing demand for such chemicals as mercury, copper, and benzene in other industries, there has been an acute shortage of these materials for manufacturing chemical pesticides. Attempts have been made to use tin, arsenic, and other organics as substitutes.[30]

The quality of chemical pesticides produced in the PRC apparently is not satisfactory. Some are not processed in accordance with the specifications, and there is no strict inspection of the products. Instructions for the use of such highly poisonous pesticides as "1605" and demeton are not adequate.[31]

In the opinion of Yang Shih-hsien, president of Nankai University in Tientsin, although more and more pesticides are being produced in many provinces, what exists in China today is infinitesimal compared with the need, and the quality of products lags far behind international standards. He suggests that knowledge of pesticides should be propagated among farmers and that special teams should be organized to handle highly poisonous and expensive pesticides in order to decrease fatal accidents and avoid wasting the pesticides.[32] Although figures for accidents caused by poisonous pesticides have never been published, the matter is apparently quite serious, as evidenced by the inquiries and discussions printed in the newspapers.[33]

In accordance with the "walking on two legs" policy, more than 500 varieties of plants and minerals have been used to make "native pesticides" for the control of plant diseases and insect pests, in addition to chemical pesticides. In 1958, 240 factories produced native pesticides and some 17 million tons of such chemicals were used. For example, extracts from tomato (Lycopersicum esculentum [Linn.] Mill) leaves have been used to cure wheat rust, and the skin of the pomegranate (Punica granatum Linn.) has been used to cure cabbage rot. Both are reported to have accomplished good results. In some areas rice blast, rice borers, wheat rust, wheat mildew, wheat smut, cotton bollworms, cotton aphids, and red spiders are reported to have been brought under control partly through the use of native pesticides.[34]

As a step to strengthen the plant protection program during the Leap Forward campaign, the Scientific Research Office for Native Pesticides was created in 1958 by ten related government agencies and research institutions. The office in turn appointed a special committee to prepare a book on native materials selected by Chinese scientists and peasants for the control of plant diseases and insect pests. The book, entitled Chung-kuo T'u-nung-yao Chih (Manual of Chinese Native Pesticides), was published in 1959 by the Science Press, Peking, and covered 522 items (403 plants and 119 minerals and other materials) that are considered effective for plant protection.

Native pesticides are said to have several advantages over chemical pesticides. China has abundant resources of plants and other materials for making native pesticides. The method of production is simple and the cost is lower. Furthermore, many of the native pesticides can be applied widely and safely to a great variety of crops. On the other hand, the available supply of some materials for making the native pesticides is seasonal and limited to certain localities. Another disadvantage is that more labor is needed in the production and application of native pesticides than for chemical pesticides. Regardless of these advantages and disadvantages, the fact is that much more native pesticide than chemical pesticide has been used in the PRC in the control of plant diseases and insect pests. In 1958, 80 percent of pesticides used were of native varieties, while only 20 percent were made of chemicals. It is likely that the same situation will remain for some time to come, mainly because of the shortage of chemicals.[35]

Many modern and native types of equipment for applying pesticides are reported to have been invented or improved. In 1966 more than 20 types of spraying and dusting equipment were in production,[36] most of them semimechanical. But the sprayers, dusters, and other appliances produced generally were of low quality and efficiency, and they were prone to break down.[37] Moreover, most peasants had very little knowledge of how to use, maintain, and repair pesticide equipment.

In 1951 the PRC first used aircraft to apply pesticides for the control of locusts in Huanghua, Hopei Province.[38] It was claimed that the area of aerial control of locusts increased from 13.7 percent of the locust-infested areas in 1958 to more than 54 percent in 1960. But the use of airplanes for applying pesticides has proved too expensive for the government. Vice-Minister of Agriculture Yang Hsien-tung has suggested that such use of airplanes should be confined to large tracts of wasteland or marshy land where the population is sparse. Except in heavily stricken areas the local governments were urged to shoulder the expenses for the use of airplanes in controlling plant diseases and insects.[39]

PROGNOSTIC AND WARNING SERVICES

During the earlier years of the Communist regime, prognostic and warning measures were taken in various areas against locusts, wheat aphids, late potato blight, rice borers, cotton aphids, and other major plant insects and diseases. But the progress was piecemeal and the methods adopted were mostly unscientific. It was not until December 1955 that a nationwide Program of Prognostic and Warning Services Against Crop Insects and Diseases was adopted. In February and March 1956 a special training class on prognostic and warning services was held in Peking under the auspices of the Ministry of Agriculture, and 184 trainees were later assigned to various provinces. At the same time a prognostic and warning section was set up in the Plant Protection Bureau of the Ministry of Agriculture and a special study team on this problem was organized by the Chinese Academy of Sciences, the Peking University of Agriculture, the North China Agricultural Science Research Institute, and the Preparatory Committee for the Chinese Academy of Agricultural Science. It was claimed that by June 30, 1956, a total of 145 prognostic and warning stations and 862 crop disease and insect information posts had been established by the agricultural experimental stations and the state farms, and that by the same date 19 prognostic and warning service research offices had been set up by various provincial governments.

At the end of 1958, the number of prognostic and warning stations was reported to have increased to 678, and that of crop disease and insect information posts to 12,000. By 1958 more than 700,000 peasants had been trained to serve as a major labor force for these stations and posts.[40]

PLANT QUARANTINE

Since 1949 the Ministry of Agriculture has issued a number of quarantine regulations and lists of objects shipped within the country and from abroad that are subject to the regulations. By the end of 1958, more than 60 quarantine offices had been established for the inspection of agricultural products and for research on methods of inspecting and sterilizing plants.

On January 1, 1958, the "Experimental Provisions for Domestic Plant Quarantine" were put into effect. These provisions were prepared by the Ministry of Agriculture in collaboration with the ministries of forestry, food, railroads, communications, and posts and telecommunications. The provisions included a list of more than 30 plant diseases and insects commonly found in China. The list was prepared on the basis of several lengthy discussions among Chinese and Soviet scientists. Seeds, seedlings, tubers, and other parts of principal crops shipped within the country were subject to inspection for these diseases and insects.[41]

The PRC has reached agreements on plant quarantine with most of the countries of the Soviet bloc. In August 1956 the Eighth International Conference on Plant Quarantine and Plant Protection was held in Peking and was attended by delegates from the Soviet Union, Poland, East Germany, Czechoslovakia, Hungary, Bulgaria, Albania, Romania, North Korea, Mongolia, North Vietnam, and the PRC. A 492-page volume containing nine reports submitted by the Chinese delegation and Chinese translations of 41 reports by other delegations was published in 1957 by the Finance and Economics Press, Peking, under the title Ti-pa-chieh Kuo-chi Chih-wu Chien-i chi Chih-wu Pao-hu Hui-i Wen-chien Hsüan-chi [Selected Documents of the Eighth International Conference on Plant Protection].

One of the major problems of quarantine work in China today is the lack of coordination. Several government agencies, including the ministries of agriculture, forestry, land reclamation, foreign affairs, and foreign trade, all have a share in the plant quarantine operations. This inevitably causes duplication, overlapping, and confusion. Also, many aspects of the quarantine regulations and the lists of plants subject to quarantine are out-of-date and have not been revised or amended in accordance with current scientific and economic developments. Finally, the techniques and facilities of plant sterilization and inspection are still below the desired standards.[42]

ENVIRONMENTAL AND BIOLOGICAL CONTROL

Plant protection measures taken by the PRC, in addition to the application of pesticides and the operation of prognostic, warning, and quarantine services, include the selection of plant varieties that are resistant to diseases and insect pests, the adoption of agriculture techniques that tend to retard the growth of germs and insects, and the use of "natural enemies" of bacteria and insects to protect crops.

By using varieties that are resistant to diseases and insect pests, considerable pesticide and manpower can be saved. It has been reported that as a result of the popularization of the Cooperation Nos. 1-5 and Kansu No. 96 varieties of wheat, both of which are highly resistant to stalk rust, this wheat disease has been basically brought under control in northeast China. Of course the breeding of varieties resistant to diseases and insect pests is not without difficulties, because a single variety cannot always combine resisting power and high yield. Also, a variety that can resist one kind of disease or insect pest often is unable to resist another. Moreover, some varieties can retain their resisting power over a long period of time, as in the case of the Nanta 2419 variety of wheat, whose resisting power to leaf rust does not degenerate for 20 to 30 years, while the resisting power of other varieties degenerates rapidly. Resisting power degenerates because the bacteria and destructive insects themselves change continuously. Therefore it is necessary to conduct intensive research on the variation of diseases and insect pests and to continually breed new varieties with stronger resisting power.

Considerable pesticide and manpower also can be saved by the adoption of certain agricultural techniques to make environmental conditions favorable to the growth of crops but unfavorable to the existence and development of bacteria, destructive insects, and weeds. For example, in some localities in Anhwei and Hunan where double-crop rice is grown, the adoption of winter plowing and winter flooding to kill the borers hidden in the rice stubble has made it possible to alleviate the damage done by rice-stem borers at the early stage. Similarly, rational crop rotation can be utilized to control cotton wilt. To a certain extent the utilization of proper cultivation methods to reduce soil temperature can alleviate the degeneration of potatoes. The proper control of water and fertilizer supply can alleviate rice blast. Sowing in straight rows and at the right time in ridge culture, coupled with the taking of preventive measures against insects, can guard against cabbage rot. However, certain localities cannot adopt such measures because of the local arrangement of manpower and other difficulties.[43]

China has a long history in the use of insects to destroy insects that are harmful to crops. Before 1600, for example, farmers in the south had begun to use Oecophylla smaragdina, a kind of ant, to control insects harmful to citrus trees. Many varieties of other insects are used for similar purposes.[44]

There are many natural enemies of bacteria and destructive insects. Some of these natural enemies can control the outbreak of diseases and insect pests over a protracted period and are not expensive to breed. In 1958-59, for example, Trichogramma evanescens (bees with red eyes) were used in Kwangtung to control sugarcane borers with good results. In Chekiang, Hupeh, and Szechuan, scarlet ladybugs (Rodolia rufopilosa Muls.) were bred on a large scale over large areas for the destruction of certain beetles. Hupeh and other provinces in central China also made successful use of small golden-green bees (Dibrachys cavus Walk.) to control the pink cotton bollworms in cotton depots. This tiny bee is a born enemy of the cotton pest. It hunts dormant pink bollworm cocoons, stings the nymphs to death, and deposits its eggs on the dead worms, which serve as food for its larvae. In 1955 two types of ladybugs (Rodolia cardinalis Muls. and Cryptolaemus montrouzieri Muls.) were introduced from the Soviet Union for the control of certain beetles in Canton and Tienpai hsien in Kwangtung, with remarkable success.[45] In the summer of 1963, the Ministry of Agriculture launched a campaign for the protection of frogs, which are considered to play an important role in exterminating harmful insects in the fields.[46]

Since 1955 horticultural workers in Szechuan have introduced red ladybugs and vedalia ladybugs that feed on orchard pests. Two ladybug-breeding bases have been set up to supply the orchards. Since the mid-1960s cotton cushion scale damage in the province has been brought under control. Bacteria also have been applied to restrain plant diseases and harmful insects. In some areas Kasugamycin, Balsticidin, and other pesticides are employed to control rice and wheat diseases. In other places initial successes have been achieved in using borer-killing bacilli to wipe out pests harmful to vegetables.

In addition scientists in Szechuan have conducted studies on the varieties, distributions, and utilization values of natural enemies of insects harmful to the major crops in the province. According to the provincial plant protection program, more biological control research groups will be set up in the rural areas.[47]

But control by biological methods also has its shortcomings. It is not as simple and convenient as chemical control, and it is less broad in the scope of application. Biological control also is affected to a greater extent by the environment, and it produces results at a slower rate.[48]

CONTROL OF HARMFUL ANIMALS AND WEEDS

Rodents are the principal animals harmful to crops. In some localities in northeast and northwest China, there are an average of about four moles to each mou of land. The number has reached as many as 30 moles to a mou in a few areas. Each mole consumes about 20 to 30 chin (27 to 40 pounds) of grain per year. With four or five moles to each mou of land, the loss of grain amounts to from 30 to 50 percent. The use of poisonous bait containing ANTU (alphanapthylthiourea) and zinc phosphide has been popularized for the control of rodents. Special efforts have been made to control rodents in spring and autumn, when their activities of consuming and storing grains are heaviest. In the spring and autumn of 1958, for instance, a total of 17 million field rats were exterminated in Kwangtung, and grain damage by rodents was basically controlled in 42 hsien of that province.[49] In 1964 and 1965 more than 2 million hectares of farmland and grassland in Chinghai were freed of rats.[50] Considerable attention also has been given to research on moles and other types of rodents, especially those in northern Shensi, Chinghai, and other parts of north China.

In the mountainous area of northern Kwantung, where mountain hogs have done considerable harm to agricultural production over a large area of farmland, since early 1970 "three-way alliance" hunting teams of militiamen, hunters, and cadres have been organized to control harmful animals.[51]

In China's rural areas weeding used to consume about one-third of the total farm labor. The menace of weeds to agricultural production was usually serious in areas that were sparsely populated, such as many localities in Heilungkiang Province. In the Mishan area of Kirin Province, *sanling* weed and darnel (*Lolium temulentum*) were found in the paddy fields. Before 1949 some previously cultivated land was left uncultivated because of these weeds, and the peasants were forced to migrate elsewhere.

In the 1950s and the 1960s wide use was made of herbicides manufactured abroad, including 2,4-D, 2,4,5-T, and IPC. In recent years considerable progress has been made in research, production, and use of herbicides. In 1972 herbicides of superior quality, such as "weeding ether," were produced in more than ten provinces and municipalities. Some herbicides manufactured from industrial by-products have been popularized. They include sha tsao an for controlling weeds in soybean fields, ti pai for controlling darnel in rice paddies, and ching yen ling for controlling wild oats, a main weed in wheat fields in northwest, northeast and southeast China.[52]

THE PROSPECTS

In the 1950s and 1960s the PRC made considerable progress in plant protection, mainly by utilizing local manpower and resources. However, diseases and insects have taken an enormous toll of growing crops and stored agricultural products, as press reports have indicated. Aside from financial and technical difficulties, mismanagement has resulted in the ineffectiveness of many programs. Substantial quantities of food and other farm products could be saved with technical and managerial improvement.

The use of modern pesticides in the PRC has thus far been restricted to state farms and large communes. It will take some time before most communes can be supplied with the funds and technical knowledge required for the utilization of modern chemicals and equipment. For this reason the PRC will have to continue the combined use of chemical and less sophisticated methods (mechanical, physical, biological, and cultural) for some time to come.

NOTES

1. Agricultural Research Service, U. S. Department of Agriculture, Losses in Agriculture, Agricultural Handbook No. 291 (Washington, D. C.: U. S. Government Printing Office, 1965), pp. 5, 41, 56.

2. J. Lossing Buck, Land Utilization in China (Nanking: University of Nanking, 1937), p. 4.

3. Shen Ch'i-i, "The Role of Plant Protection in Insuring the Increase of Agricultural Production," Hung Ch'i [Red Flag], Peking, No. 22, 1962, pp. 33-41.

4. Bureau of Plant Protection, Ministry of Agriculture, People's Republic of China, "Stir up the High Tide for Thorough Control of Plant Diseases and Insect Pests," Chung-kuo Nung-pao Tseng-k'an [Supplement to Chinese Agricultural Bulletin], Peking, No. 3, 1958, pp. 1, 4.

5. T'an Chen-lin, "Strive for the Fulfillment, Ahead of Schedule, of the National Program for Agricultural Development," JMJP, April 7, 1960, p. 1.

6. Yang Hsien-tung, "Strive for Greater Success on the Plant Protection Front," Chung-kuo Nung-pao [Chinese Agricultural Bulletin], Peking, No. 6, 1963, pp. 1-4.

7. Kao Min-chen, "Great Achievements in China's Plant Protection Work in the Past Decade," Chung-kuo Nung-pao, No. 15, 1959, pp. 24-26.

8. Jen-min Shou-ts'e [People's Handbook] (Peking, 1952), pp. 375-77.

9. Bureau of Plant Protection, Ministry of Agriculture, People's Republic of China, Ti-pa-chieh Kuo-chi Chih-wu Chien-i chi Chih-wu Pao-hu Hui-i Wen-chien Hsuan-chi [Selected Documents of the Eighth International Conference on Plant Quarantine and Plant Protection] (Peking, 1957), pp. 37, 41-42; Shen Ch-i-i, op. cit.

10. NCNA, Peking, December 8, 1962.

11. Yang Hsien-tung, op. cit.

12. NCNA-English, Peking, July 16, 1963.

13. Research Institute of Plant Protection, Chinese Academy of Agricultural Science, "Accomplishments of Plant Protection in China During the Past Decade," Nung-yeh K'o-hsueh T'ung-hsun [Agricultural Science Bulletin], Peking, No. 18, 1959, pp. 629-32; Shen Ch'i-i, op. cit.

14. NCNA-English, Peking, November 8, 1973.

15. "War Against Rice Borers Must Be Won This Year," editorial, JMJP, May 10, 1964, p. 1.

16. Bureau of Plant Protection, Ministry of Agriculture, People's Republic of China, op. cit., pp. 48, 474-75.

17. Ibid., pp. 43-44; Tai Sung-en, "Fully Develop the Role of Crops of Good Breed to Increase Production," Hung Ch'i, No. 2, 1963, pp. 25-34.

18. Shen Ch'i-i, op. cit.

19. Research Institute of Plant Protection, Chinese Academy of Agricultural Science, op. cit.

20. "Breeding Good Strains," Peking Review, No. 7, 1973, pp. 16-18.

21. NCNA, Peking, September 23, 1960.

22. Bureau of Plant Protection, Ministry of Agriculture, People's Republic of China, op. cit., pp. 45-46.

23. Ibid., pp. 47-48.

24. Research Institute of Plant Protection, Chinese Academy of Agricultural Science, op. cit.

25. Bureau of Plant Protection, Ministry of Agriculture, People's Republic of China, op. cit., pp. 44-45.

26. Shen Tsung-han, Agricultural Resources of China (Ithaca, N.Y.: Cornell University Press, 1951), p. 52.

27. NCNA-English, Peking, March 8, 1961.

28. For comprehensive coverage of the current situation on pesticides in the PRC, see Fei Yu-ch'un, ed., Nung-yao Wen-ta [Questions and Answers on Agricultural Chemicals] (rev. ed.; Peking: Fuel Chemical Industry Press, 1975).

29. NCNA-English, Peking, October 9, 1973.

30. Chinese Academy of Sciences, "Organic Tin and Organic Chemistry," K'o-hsueh T'ung-pao [Scientia], Peking, No. 19, 1959, pp. 653-54.

31. Shen Ch'i-i, op. cit.

32. Yang Shih-hsien, "A chat About the Pesticide Problem," JMJP, December 4, 1962, p. 5.

33. Yao Sun-ch'ien, "Attention Must Be Given to Safety When Using Poisonous Pesticides," JMJP, June 9, 1962, p. 2.

34. Research Institute of Plant Protection, Chinese Academy of Agricultural Science, op. cit.

35. P'ei Li-sheng, "Ten Big Advantages in Popularizing Native Pesticides," K'un-ch'ung Chih-shih [Knowledge of Insects], Peking, No. 2, 1959, pp. 49-55.

36. NCNA-English, Peking, February 26, 1966.

37. Shen Ch'i-i, op. cit.

38. Jen-min Shou-ts'e, loc. cit.

39. Yang Hsien-tung, op. cit.

40. Kao Min-chen, op. cit.

41. Plant Quarantine Laboratory, Ministry of Agriculture, People's Republic of China, Tui-nei Chih-wu Chien-i Tui-hsiang T'u-shuo [Illustrated Explanations on the Objects for Domestic Plant Quarantine] (Peking, 1956), p. 3; "Experimental Provisions for Domestic Plant Quarantine," Nung-yeh K'o-hsueh T'ung-hsun, No. 1, 1958, pp. 17-20.

42. Chen Shan-min, "Certain Problems in Plant Protection Work," JMJP, December 4, 1962, p. 5.

43. Shen Ch'i-i, op. cit.

44. Hsu Wei-liang, "Insect-Eating Insects," Sheng-wu-hsueh T'ung-pao [Biological Bulletin], Peking, No. 2, 1966, pp. 17-21.

45. Chao Shan-huan, "China's Accomplishments in the Control of Insects Harmful to Agriculture in the Past Ten Years," K'un-ch'ung Chih-shih, Peking, No. 8, 1959, pp. 241-43.

46. Yang Hsien-tung, "Protecting Frogs," JMJP, September 12, 1963, p. 2; Cheng Tso-hsin, "Frogs' Functions in Protecting Agriculture," JMJP, September 12, 1963, p. 2.

47. NCNA-English, Peking, March 14, 1975.

48. Shen Ch'i-i, op. cit.

49. Kao Min-chen, op. cit.

50. NCNA-English, Sining, January 4, 1966.

51. NCNA, Canton, August 22, 1970.

52. Peking Review, No. 26, 1972, p. 21.

CHAPTER

17

TOOL IMPROVEMENT

Farm operations in China are accomplished largely by human labor, with very little or no assistance from animals and machinery. Work thus done is slow and tedious. Frequently crops either become overripe or are damaged by unfavorable weather conditions before they can be harvested. Despite its large population China actually suffers from a shortage of farm labor during busy seasons, and the employment of extra hands at harvesting time is virtually impossible. There is an urgent need for the improvement of agricultural implements.

Because of the low purchasing power and low educational level of the farmers, the introduction of new and improved implements is difficult without extensive government assistance. For decades prior to the Communist take-over, the national and provincial authorities as well as private institutions (including the University of Nanking, Ling Nan University, and Southeastern University) attempted to introduce and popularize agricultural implements and machinery from abroad. But the influence of the imported implements and machinery was limited. The establishment of the National Agricultural Engineering Corporation in 1943 by the Ministry of Agriculture and Forestry, together with the Farmers' Bank, represented the first attempt on the part of the government to place the manufacture of farm implements and machinery on a national and organized scale. The program called for the establishment in the principal agricultural centers of plants to manufacture implements and for a plan of education and demonstration. However, the accomplishments were sporadic and small-scale.[1]

The general policy at the outset of the Communist regime was to improve farm implements and to move gradually toward mechanization. During the Economic Recovery Period (1949-52) efforts were made to increase supplies of old-fashioned implements and to popularize some new implements through demonstrations. In the First Five-Year Plan period (1953-57) many innovations of farm implements were developed and some 5.11 million improved and modern farm implements for plowing, tilling, raking, pressing, sowing, harvesting, and threshing were made available to farmers.[2] During the Leap Forward period an ambitious campaign was launched to speed up both agricultural mechanization and tool innovation in accordance with the "walking on two legs" policy, but the campaign was not very successful. In recent years the policy has been simultaneous development of mechanization and semimechanization, with the main stress on the latter

IMPROVEMENT OF OLD-STYLE IMPLEMENTS

Article 7 of the NPAD provided:

> Energetic efforts should be made to extend the use of improved and new types of farm tools suited to local conditions on the basis of the needs of production and by trial and remodeling to adapt them to the conditions of the particular region. Attention should always be paid to the maintenance and repair of farm tools. Technical guidance in the use of the new types of farm tools should be strengthened.

Perhaps the most important item under the program for the improvement and innovation of old-fashioned agricultural implements was the plan for the popularization of some 6 million double-wheel/double-share plows and double-wheel/single-share plows throughout the country within three to five years, beginning in 1956. Both types were animal-drawn and regarded as suitable for cultivating sandy or light clay soils, and the single-share type also was regarded as suitable for reclaiming wasteland.[3] But this plan was almost a complete failure. The plows were mainly copies of those used in the broad, flat wheat lands of the Soviet Union and were unsuited to China's terraced and small plots. Many farmers complained that the plows were too heavy, that their farm animals were too few and too weak to pull the plows, and that they had no means of repairing the plows even if they had been able to use them successfully. In south China farmers complained that the rice paddies were too small and too muddy for the plows. Consequently the plan was revised to include only 3.5 million plows, and finally only 100,000 to 150,000 were distributed to farmers.[4] It was reported that in 1957 only 72,000 double-wheel/double-share plows were distributed to farmers, while 800,000 were left idle, and that many of those distributed were later returned or destroyed by farmers.[5]

At the beginning of the Leap Forward campaign in 1958, a further effort was made to popularize the double-wheel/double-share plows in areas other than the rice fields of south China.[6] It also was reported that a modified type, which eliminated the drawbacks of the original type, had been designed by the Chekiang Agricultural Research Institute in Hangchou and was personally tried by Mao Tse-tung during his visit to the institute in January 1958.[7]

Since rice is the most important crop in China and transplanting of seedlings is the most labor-consuming operation in rice growing, the PRC has given considerable attention to the improvement of rice transplanters of all types—those operated by hand, those operated by animals, and mechanized transplanters—although manual types remain the most widely used. The first improved model was introduced in 1956. It was an oxen-drawn transplanter developed jointly by seven agricultural research institutes, and it was reported to be able to do the work of 30 persons.[8] During the Leap Forward campaign the use of locally constructed transplanters of wood and bamboo was extensively publicized.

The number of rice transplanters of various kinds in use throughout the country increased from 31,000 in 1966 to 230,000 in 1972 and to 277,000 in 1973.[9] One type of rice transplanter developed during this period, said to be able to transplant seedlings with soil at the root, is now widely used in south China.[10]

In the 1960s the Kwangsi-Chuang Autonomous Region led in the use of rice transplanters. In 1968 the region had more than 30,000 transplanters, mostly of the Kwangsi-65 hand-propelled type, working over 53,000 hectares of paddy fields.[11]

Many farm tools and implements in addition to plows and rice transplanters were reported to have been developed or innovated by the masses during the Leap Forward period, for use in manure accumulation, transport, irrigation and drainage, tilling, processing of farm products, and other purposes.[12] But many of these tools and implements were lost, wasted, or destroyed, mainly because of mismanagement and the overemphasis on mechanization. As reported by the People's Daily, "They were left scattered in the open air in the field where rains and winds ruined them."[13] It was not until the latter part of 1960 that responsibilities toward farm tools and implements were clearly defined for the commune and its components (the production brigade, the production team, and the individual commune member). After the country had suffered two consecutive years of bad harvest in 1959-60, the policy of "taking agriculture as the foundation of the national economy" was adopted. A movement was started in August 1960 to send all available hands to the fields, and all available land was devoted to grain production. In order that these people would not go to the fields empty-handed, the importance of medium-size and small tools and implements was again emphasized.[14]

Under the new directives such medium-size implements as plows, harrows, weeding disks, waterwheels, carts, fertilizer dispensers, and threshers should be under the care of the commune or of the production brigade or production team using them. A shed should be built to store them. A commune member "known for his high consciousness and his zeal for common property" should be appointed guard, and he should also inspect these implements and see that they are repaired in good time. Individual commune members were responsible for the care and use of small implements (such as spades, hoes, picks, sickles).[15] In all cases a pressing duty should be "the education of the masses toward love of the common property."[16]

It was further admitted that care of the existing implements would not alone be sufficient to provide for the millions who had been urged to take part in agricultural production. There was an urgent need for the manufacture of new implements as well as the repair of old ones. Survey reports from a number of communes revealed almost uniformly that the shortage of farm implements was very seriously felt.[17]

But manufacturing and repair were impeded by an acute shortage of steel and other necessary materials. Although it was reported that in 1962 the state allocated and made arrangements for the delivery of some 2 million cubic meters of lumber and "corresponding quantities" of rolled steel, copper, and hemp to various localities for the repair, maintenance, and manufacturing of farm implements, many hsien had to obtain much of the material locally, to use various substitutes, and to collect scrap iron, pieces of wood, split bamboo, nails, and screws. The process of casting, instead of forging, had to be adopted for making implements and spare parts because the "native" iron cannot be forged; the use of cheap native iron and steel resulted in the breakdown and deterioration of millions of implements produced or repaired in innumerable improvised small factories.[18] In the 1960s the supply and maintenance of small and medium-size farm implements as a whole seemed to have improved somewhat; but the problems of high costs, high depreciation rates, shortage of necessary materials, and poor management remained unsolved in many areas.

MECHANIZATION OF FARMING

Although the Chinese Communists realize that agriculture in China can be mechanized only gradually, in view of limited industrial capacity and raw materials, nationwide farm mechanization is one of the great benefits that the Communist regime promised to bring to the peasants from the outset. Mechanization was first emphasized during the land reform in 1951-52, and again during the cooperativization of agriculture in 1955-56. But prior to the Leap Forward campaign the PRC's plan for agricultural mechanization was by no means very ambitious. Thus in a report delivered on July 31, 1955, at a meeting of the secretaries of the provincial, municipal, and district committees of the CCP, Mao Tse-tung stated: "All through the First and Second Five-Year Plans, the number of big farm machines will certainly increase, but not to any great extent. During the Third Five-Year Plan, more big farm machinery will be employed year by year." He estimated that it would take four or five five-year plans to complete agricultural mechanization in China.[19]

The provision for agricultural mechanization in Article 7 of the NPAD likewise seemed moderate:

> Keeping pace with the development of industrialization of the country, mechanization of agriculture should be carried out energetically step by step. Within the shortest possible time beginning in 1956, the engineering and agriculture departments should draw up, on the basis of extensive experiments and research, a program for agricultural mechanization suited to the conditions in China, so as to make good preparations to extend the mechanization of agriculture. Then machinery suited for use in various places should be improved as circumstances require and its use extended energetically.

In any case agricultural mechanization did not make spectacular progress during the Economic Recovery Period and the First Five-Year Plan period, although it was reported that the number of tractors increased from 401 in 1949 to 24,629 in 1957 and that the number of other agricultural machines also increased considerably.[20] It was not until the launching of the Leap Forward campaign in 1958 that an ambitious program of agricultural mechanization was mapped out. In order to expedite farm mechanization, the State Council set up the Ministry of Agricultural Machine-Building Industry in August 1959 (renamed the Eighth Ministry of Machine-Building Industry in December 1964), with the functions of administering the manufacture and popularization of tractors and other farm machines.[21] Meanwhile, China's first large tractor plant went into full production in November 1959 at Loyang, Honan Province. It was built with technical aid from the Soviet Union and was estimated to have an annual capacity of 15,000 54-horsepower tractors. Additional tractor plants have since been constructed in Tientsin, Nanchang, Changchun, Anshan, Shenyang, and a few other cities. A number of factories manufacturing other agricultural machines also have been established or expanded.

According to Ch'en Cheng-jen, minister of agricultural machine-building industry, the general aim was to enable agriculture, forestry, animal husbandry, fishery, and side occupations to advance gradually toward overall mechanization and electrification within about ten years. However, owing to insufficient production of machinery, primary attention would be given to agriculture and animal husbandry during the first four years, striving first to mechanize a small part of agricultural and pastoral production and a fundamental portion of farmland irrigation. During this period emphasis would be placed on the suburbs of the large cities, the commodity-grain production centers, the major centers of other economic crops, and the principal pastoral areas. Ch'en Cheng-jen estimated that the supply of farm machinery would be adequate to permit completion of over half the task of mechanization in seven years and to complete it fully in ten years.[22]

But the progress of agricultural mechanization during and immediately after the Leap Forward campaign (1958-62) was considerably slower than had been expected. At the end of 1960, only 5 percent of the country's cultivated land was cultivated with machines (the target for that year was 6.9 percent), 15 percent with semimechanized methods, and 80 percent with improved small implements.[23] Toward the end of 1962, the area farmed by tractors still accounted for less than 10 percent of the cultivated land.[24]

The degree of agricultural mechanization in 1971 was described as follows: "number of tractors, double the 1962 figure; irrigation and drainage mechinery, three times as much as in 1962; amount of electricity consumption in the rural areas, six times as much as in 1962; and machinery for forestry, animal husbandry, side-line occupations and fisheries, fairly big increase." It also was reported that in 1971 tractors and harvesters were used on vast areas of farmland on the northeast China plain, and that mechanized and electric irrigation and drainage had been popularized on the north China plain. In the major rice-producing areas in south China, rice transplanters and other mechines for use in paddy fields were being popularized on large areas, while advanced machinery for animal husbandry and fisheries had been adopted in the frontier grazing lands and the southeast coastal fishing grounds.[25]

Mechanization materialized primarily on the state farms and some large communes. In Heilungkiang Province, where many large state farms are located, 20 percent of the cultivated land was cultivated in 1960 by machines and more than 30 percent by modern animal-drawn implements.[26] In 1970, 40 percent of the province's cultivated land was farmed mechanically and about 50 percent of the production brigades in the province used machines in threshing, milling, grass-cutting, and grinding. They were considered as "having fundamentally achieved agricultural mechanization."[27] Among other areas where the acreage of mechanized farming made up an important portion of the total area of cultivated land were Hopei and Shantung provinces, the Sinkiang-Uighur Autonomous Region, and the suburbs of Peking and Shanghai.

Although the number of tractors in operation was reported to have increased rather rapidly, from 401 in 1949 to 770,805 in 1972 (see Table 17.1), the increase was far from sufficient to meet the current needs. As the People's Daily pointed out, on the basis of one tractor for 1,500 mou (250 acres), more than 1 million tractors would be required for the mechanization of the estimated 1.6 billion mou (267 million acres) of farmland in China.[28]

Of the 45,330 tractors in operation in 1958, 957 were produced in China.[29] Production figures for more recent years are not available, but it was claimed that by the end of 1965, the country had developed 13 types of tractors, including heavy (100-horsepower) and medium-size caterpillar tractors for large-scale land reclamation and contouring, medium-size wheeled tractors for haulage of timber as well as ordinary fieldwork, light tractors (under six-horsepower) for garden-type cultivation, and tractors specifically designed for rice paddies.[30]

In addition to tractors, the PRC was reported to have developed a wide range of agricultural machinery and semimechanized implements for irrigation and drainage, sowing, harvesting, transplanting, fertilizing, controlling weeds, and processing agricultural products. For example, most of the new farm machines and semimechanized tools used on the 36 state farms in the Sanchiang Plain Reclamation Area in Heilungkiang Province in 1965 were made in China, including tractors from Loyang, combine harvesters from Kaifeng, tractor-drawn five-share plows from Anshan, cultivators from Tsitsihar, and sowing machines from Sian.[31]

Since the Leap Forward campaign the agricultural machinery industry in the PRC has placed increasing emphasis on mechanized pumps. In 1962 the country had 26 large factories producing pumps and power units for pumps.[32] Some of the factories originally built to manufacture tractors were producing pumps as well as tractor parts, because pumps require much less steel, capital, and fuel than tractors and because mechanization of irrigation and drainage is considered more urgent than that of land cultivation.

Mechanized wheat cultivation has been introduced in most parts of the principal wheat-producing areas—northeast China, the Yellow River basin, and the Huai River basin. The East Wind

TABLE 17.1

Number of Tractors in the PRC
1949-65, 1970-72

Year	Number of 15-Horsepower Units
1949	401
1950	1,286
1951	1,410
1952	2,006
1953	2,719
1954	5,061
1955	8,094
1956	19,367
1957	24,629
1958	45,330
1959	59,000
1960	79,000
1961	99,000
1962	100,000
1963	115,017
1964	123,000
1965	154,161
1970	308,322
1971	700,732
1972	770,805

Sources: For 1949-58, State Statistical Bureau, Ten Great Years (English ed.; Peking, 1960), p. 135; for 1959, Ch'en Cheng-jen, "Speed up the Technical Transformation of Agriculture," Hung Ch'i [Red Flag], Peking, No. 4, 1960, pp. 4-10; for 1960 and 1961, Chang Feng-shih, "Produce Suitable Agricultural Machinery to Support Agriculture," Kung-jen Jih-pao [Daily Worker], Peking, January 18, 1962, p. 1; for 1962, "Actively and Systematically Bring About the Technical Transformation of our Country's Agriculture," editorial, JMJP, November 9, 1962, pp. 1-2; for 1963, news release, China News Service, Canton, January 5, 1964; for 1964, Chou En-lai, "Report on the Work of the Government," presented at the First Session of the Third National People's Congress, December 21-22, 1964, NCNA-English, Peking, December 30, 1964; for 1965, "Paving a Way to Agricultural Mechanization in China: A Visit to the Agricultural Mechanization Hall of the National Agricultural Exhibition," JMJP, April 13, 1965, p. 1; for 1970, "Some Basic Facts About the People's Communes," supplement to China Reconstructs, Peking, Vol. XXI, No. 1 (January 1972), pp. 10-13; for 1971, Economic Reporter, Hong Kong, English supplement, No. 1, 1973, p. 4; for 1972, Chung-kuo Hsin-wen (China News), Hong Kong, No. 6713, February 7, 1973, p. 1.

combine harvester, which first appeared in 1964, reaps about 13.3 hectares of wheat per day. In northeast China complete sets of farm tools for ridge culture and a corn seeder have been popularized. For south China's paddy fields several kinds of rice transplanters and a rice harvester to be attached to the walking tractor have gone into mass production. In many parts of south China where silt from lakes and rivers is available, a wide variety of suction pumps and winches for silt extraction have been in general use. Types of apparatus for spraying liquid ammonia and for applying powder or granular fertilizers have appeared in the countryside. There also is an increasing range of machines for processing agricultural products, including machines for rice husking, flour grinding, potato slicing, cotton ginning, and oil pressing.[33]

Domestic production of tractors and other agricultural machines has been handicapped by the shortage of raw materials, particularly steel, electric power, coal, gasoline, and other fuels.[34] The few centers of the steel industry have experimented with new products suitable for agricultural machinery and more than 700 types of rolled steel were reported to have been produced in 1962, but this amount is not commensurate with the needs of the country.[35] Certain types of steel products and essential parts of agricultural machinery cannot be produced in China and therefore must be imported. The East Is Red tractor, for example, consists of some 10,000 parts requiring more than 450 kinds of steel and other metals. When the tractor was first manufactured in 1959, most of these metals were imported. Although the variety of metals produced domestically has increased, in 1962 there were still more than 30 kinds of metals needed for making farm machinery that China could not produce or could produce only in small quantities. Furthermore, the quality of some of the metals and parts produced in China was below the required standard.[36]

Other serious problems in addition to lagging production have handicapped the mechanization of agriculture. Tractors were first managed by the state-operated tractor stations, but many of them were turned over to the communes during the Leap Forward campaign. Despite the claimed advantages of collective strength and self-reliance, management under the communes has been far from satisfactory. The depreciation rate and maintenance costs have been high, partly because of "lack of the sense of political responsibility to take good care of tractors on the part of drivers." In 1961 about 20 percent of tractors and 20-30 percent of irrigation machines needed repair; many tractors were without machine-drawn farm tools, and irrigation and drainage machines were without pipes and pumps. Moreover, operators and maintenance personnel for tractors and other farm machines generally had undergone only a very short period of technical training.[37]

Until the late 1960s very little attention was paid to producing attachment tools and spare parts or to maintenance and repairs. Hsiang Nan, secretary of the Secretariat of the Chinese Young Communist League, blamed low tractor efficiency and high operational cost on the imbalance of investment in the four different segments of the industry. He pointed out that a 200-million yuan investment made on a tractor plant should be matched with additional investments of 200 million yuan on plants producing tractor-drawn tools, 200 million yuan on plants producing spare parts, and 300 million yuan on a network of maintenance and repair stations.[38]

The use of a great variety of agricultural machines has created additional problems. A visitor to the six production brigades in Shulu hsien, Hopei Province, found that as many as eight models of motors of different horsepower and consuming different fuels (natural gas, gasoline, and so on) were used for various machines. In the Ningpo district, Chekiang Province, more than 40 types of internal combustion machines were used for irrigation and drainage. This situation made it difficult for the technicians to handle the machines because most of them had had only limited technical training. It also created the problem of getting an adequate supply of all kinds of fuels and accessories.[39]

Another technical problem is that in many areas farmland still does not meet the conditions for satisfactory mechanized farming. In Pingku hsien of the municipality of Peking, for example, the local authorities instructed in 1962 that land less than 10 mou (1.7 acres) in area or less than 150 meters (165 yards) in length should not be cultivated with caterpillar tractors, and that land less than 5 mou (0.85 acre) in area or less than 100 meters (110 yards) in length should not be cultivated with wheeled tractors. Under these instructions there is very little land suitable for mechanical cultivation. Moreover, even after the agricultural collectivization, many parcels of land still have not been properly adjusted to make them suitable for mechanization. Within each piece of land there are abandoned wells and ditches not yet flattened or boundary stones not yet removed. Also, on some pieces of land, irrigation canals are not well located and sometimes divide the land into a number of small fragments so that mechanization of farming can hardly be carried out. Finally, some roads, bridges, and tunnels between the farms are not sufficiently wide or strong for the passage of large farm machines, which sometimes must take a long, roundabout route from one place to another.[40]

Still another problem that has handicapped agricultural mechanization in the PRC is that under the three-level ownership system, the production team, as the basic accounting unit of the commune, has no ability to buy large quantities of farm machinery and is not suitable for centralized operation of large machines.[41]

For years the problem of how agricultural mechanization should be carried out was heatedly debated between Mao Tse-tung and Liu Shao-ch'i, although the differences in their opinions were not revealed until the Cultural Revolution. The focus of the dispute was the question of "who should run agricultural mechanization." In Mao's opinion the collective strength of peasants should be relied on as the mainstay of agricultural mechanization. The communes or the production brigades should have the full ownership of farm machines and power to operate them. It was Mao's belief that the communes or the production brigades could achieve agricultural mechanization with greater, faster, better, and more economical results. For this reason cooperativization should precede agricultural mechanization. Liu Shao-ch'i, on the other hand, was of the opinion that the communes or the production brigades did not have sufficient experience or adequate funds to run agricultural mechanization, and that the state should first invest funds and train technical personnel. Later, agricultural mechanization might gradually be taken over by the communes.

In November 1962 T'an Chen-lin, director of the State Council's Office of Agriculture and Forestry, in "Opinions on Readjusting and Improving the Work of Tractor Stations," asserted that the communes "have no means to purchase, manage and run" tractors, which should be handed over to the state for operation. For several years thereafter the communes no longer ran tractor stations. But during the Cultural Revolution the policy of the government on agricultural mechanization was severely criticized by the Maoists as a revisionist line of Liu Shao-ch'i. The operation of tractor stations was taken as one of the key points in dispute, and in many hsien tractor stations were again placed under the management of the communes.[42]

Despite the slow progress, even toward the end of 1960 the People's Daily still maintained that modernization of small agricultural implements would be accomplished in four years beginning in 1959, that of medium-size implements in seven years, and complete agriculture mechanization in ten years.[43] It was not until the beginning of 1962, after three consecutive years of bad harvest, that the failure of agricultural mechanization was admitted. Chang Feng-shih, vice-minister of agricultural machine-building industry, revealed:

> In the two years of 1960 and 1961, the agricultural machine-building industry has supplied agriculture with 40,000 standard tractors, 37,000 motor-powered plows, 37,000 motor-powered harrows, 3.3 million horsepower of agricultural irrigation equipment, 150,000 rubber-wheel animal propelled carts, 11.5 million rubber-wheel hand carts, 800 million small agricultural tools and many other semi-mechanized agricultural tools. But these are

far from satisfying the agricultural needs. Furthermore, the quality of certain products is not high. They are not properly coordinated, and their supply is not in time.[44]

DRAFT ANIMALS

Many medium-size farm implements depend on draft animals as a principal source of power for operation. Even in some areas with a higher degree of mechanization, draft animals are needed because of the particular types of farming and because of the limitations of climatic and soil conditions. The cultivated land of Heilungkiang, for example, is composed of black, yellowish-black, and light salty soil, which is sticky and has a comparatively low rate of evaporation. After a downpour tractors are of no use, and certain kinds of farming would be impossible without draft animals. Therefore a combined use of machines, horses, and oxen will remain essential to agricultural production.[45]

The NPAD called for the mapping out of different plans for the cooperativization of animal husbandry in accordance with local conditions, and for the development of state livestock farms. Specifically the program emphasized the protection and breeding of domestic animals and fowl; special care for female, young, and stud animals for breeding purposes; the establishment of breeding stations; and the improvement of livestock strains. For pastoral areas the program called for the protection of grassland, the improvement and cultivation of green pastures, the opening up of water sources, the gradual establishment of feed and fodder bases by livestock cooperatives, and the popularization of ensilage.

The program also required that, wherever possible, diseases most harmful to domestic animals—such as rinderpest, hog cholera, Newcastle disease, contagious pleuropneumonia of cattle, foot and mouth disease, pork measles (cysticercosis), lamb smallpox, and sheep mange—be fundamentally eradicated within a period of 7 to 12 years. By 1962 veterinary stations were to be set up in all hsien, chu, and hsiang in agricultural and pastoral areas; cooperatives were to have personnel with basic training in the prevention and treatment of animal diseases; and peasant veterinarians were to be developed, organized, and guided in order to improve their technique in veterinary work (Article 3).

The program also provided that in the 12 years beginning in 1956, every APC (or commune) should raise an adequate number of strong draft animals and that sufficient grass and fodder crops for the draft animals should be produced. The communes were urged to constantly improve the breeding and management of draft animals according to their conditions (Article 14).

The number of draft animals in the PRC increased from 59,775,000 in 1949 to 87,388,000 in 1955 but decreased steadily to 84,689,000 in 1958 (see Table 17. 2). The net gain from 1949 to 1959 was 43 percent, an average annual gain of 4. 3 percent. No official figures on draft animals have been released since 1959, but according to the Communist Chinese press the progress of draft animal production in Inner Mongolia, Sinkiang, Chinghai, Tibet, and other important livestock areas has been slow in recent years.

There were several causes for the slow growth. First, many animals weakened or died because of overwork, excessive heat or cold, and diseases. Draft animals were used in China traditionally for irrigation, for turning the primitive village mills, and for local transport. Their functions increased tremendously during the Leap Forward years, when deep plowing was introduced. However, it was not until the summer of 1960 that several provincial authorities issued orders for the strengthening of draft animals. In Heilungkiang it was directed that animals should be granted some rest and that during gestation females should be exempted from work for one month.[46]

Another reason for the failure in propagation of draft animals was the absence of incentive, because of low wages, among the herdsmen and those in charge of animals in the cooperatives or communes. Cases of deliberate neglect and secret slaughter have been reported in some areas. Under the communes there were practically no systems of care, which did exist at the time of APCs; and the men in charge of the animals were paid uniformly, regardless of whether the work was well or badly done. The need to reintroduce a regular system of management was emphasized at the end of 1960, and it was urged that production teams and persons in charge of animals be given a premium for work well done.[47] Beginning in the spring of 1961, the care and mating of animals was given over to the production brigades or production teams, and the payment to the persons in charge of animals was readjusted.[48]

The shortage of fodder was another cause of the failure to improve the situation of draft animals. An improvement in the feeding of livestock and in the related problems of management is a prerequisite to any substantial livestock improvement program through breeding. However, draft animals in China were fed primarily on roughage and received only limited amounts of concentrates, although the Chinese Communist authorities made efforts to improve the feeding of farm animals. It was reported that by the end of 1958, the chemical composition of 1,032 kinds of feed from different parts of the country had been analyzed. Of these 581 were green feeds, many

TABLE 17. 2

Number of Draft Animals in the PRC,
1949, 1952-59
(thousands)

Year	Cattle	Horses	Donkeys	Mules	Total
1949	43,936	4,874	9,494	1,471	59,775
1952	56,600	6,130	11,806	1,637	76,173
1953	60,083	6,512	12,215	1,645	80,455
1954	63,623	6,939	12,700	1,717	84,979
1955	65,951	7,312	12,402	1,723	87,388
1956	66,601	7,372	11,686	1,711	87,370
1957	65,860	7,510	10,900	1,840	86,110
1958	64,952	7,512	10,601	1,624	84,689
1959	n. a.	n. a.	n. a.	n. a.	85,400

Note: n. a. = not available.

Sources: For 1949 and 1958, Ts'ai Tsu-wei, "Accomplishments in Animal Husbandry in the Past Decade," Chung-kuo Hsu-mu-hsueh Tsa-chih [China Animal Husbandry Magazine], Peking, No. 10, 1959, pp. 289-93; for 1952-57, Yu. N. Kapelinskiy et al., Development of the Economy and Foreign Economic Contacts of the People's Republic of China Moscow, (1959), translated from the Russian by the Joint Publications Research Service, JPRS No. 3234 (Washington, D. C., May 23, 1960), p. 311; for 1959, JMJP, January 26, 1960, p. 3.

of which grow wild and have a fairly high nutritive value. Studies also have been made on methods of expanding cultivation of certain high-quality feeds in various parts of the country.[49] But despite all these efforts the supply of fodder remained insufficient. The situation was particularly serious during the three consecutive years of natural calamities in 1959-61, which caused draft animals to be badly underfed.

The propagation of draft animals also was adversely affected by the shortage of technical personnel and necessary instruments. A directive on the promotion of mating of large animals, issued in February 1959 by the central government, indicated that there were 27 million head of female animals at the right age for propagation, of which 15 million were pregnant. Artificial insemination was encouraged in order to increase the number of large animals by 10 percent by the end of 1959, but trained technicians and necessary instruments were not sufficient.

The shortage of veterinarians remained very acute in the 1960s, and the strengthening of the veterinary centers throughout the country was urged.[50] In 1965 the PRC had about 10,000 veterinarians, most of them graduates of the 17 universities with veterinary training facilities. In addition there were 20,000-30,000 graduates from a number of professional veterinary schools that offered a three-year curriculum.[51]

The "walking on two legs" policy called for the simultaneous development of native and modern veterinary methods. Traditional Chinese veterinarians had been practicing in villages for centuries, using knowledge handed down from father to son, but they had no real scientific knowledge. In a directive on the strengthening of veterinary work, issued by the State Council in January 1956, modern and native veterinarians were urged to unite and to exchange knowledge. Special efforts have since been made to collect native prescriptions and native herbs and to study native methods of diagnosis and treatment, particularly acupuncture. The Western-trained veterinarians' task usually is that of diagnosis, while most of the treatment is left to the native veterinarians.[52]

In recent years many provinces and autonomous regions have selected students from among the poor and lower-middle-class peasants and trained them as "barefoot" veterinarians. Working closely with personnel of the veterinary stations, the "barefoot" veterinarians have organized teams to work in the rural areas.

ELECTRIFICATION

Special mention should be made of one aspect of agricultural mechanization, namely electrification, because it is regarded by the Chinese Communists as one of the four major transformations (szu hua) essential to the modernization of Chinese agriculture, the others being mechanization, water conservation, and "chemicalization" (the widespread use of chemical fertilizers and chemical pesticides).

Prior to the Communist take-over the use of electricity in China's rural areas was infinitesimal. The progress of rural electrification during the Economic Recovery and the First Five-Year Plan periods was slow, mainly because of financial and technical difficulties. By the end of 1957, the country's rural areas had only 544 hydroelectric power stations with a total capacity of over 20,000 kilowatts (see Table 17.3). In fact no goals for rural electrification were provided in the NPAD, 1956-67, which merely stated in Article 5: "In building water conservation projects where power generation is possible, every effort should be made to undertake the simultaneous construction of small and medium-sized hydroelectric power stations; and by coordinating

TABLE 17. 3

Number of Hydroelectric Plants in Rural China
at the End of the Year,
1949-58, 1960, 1972-73, 1975

Year	Number of Plants	Total Capacity (kilowatts)	Average Capacity (kilowatts)
1949	57	5,330	94
1950	68	6,248	92
1951	88	7,804	89
1952	98	8,137	83
1953	105	8,321	80
1954	114	8,445	74
1955	127	8,900	70
1956	240	11,860	49
1957	544[a]	20,324[a]	37[a]
1958	4,878	151,826	31
1960	9,000[b]	250,000	27.8
1972	35,000[b]	n. a.	n. a.
1973	50,000[b]	n. a.	n. a.
1975	60,000[b]	n. a.	n. a.

[a]Based on incomplete statistics.
[b]Small hydroelectric plants only.
n. a. = not available.

Sources: For 1949-57, Shui-li Fa-tien [Hydroelectrics], Peking, No. 12, 1958, p. 54; for 1958, Chang Chung-wei et al., "Communization Is Beneficial to Rural Electrification," JMJP, January 26, 1960, p. 7; for 1960, NCNA-English, Peking, February 23, 1972; for 1972, Chiang Lei, "Energetically Develop Small Hydroelectric Power Stations in the Countryside," Hung Ch'i [Red Flag], Peking, No. 3, 1972, pp. 78-81, NCNA-English, Peking, September 28, 1973; for 1973, NCNA Domestic Service in Chinese, Peking, January 3, 1974; for 1975, NCNA-English, Peking, September 10, 1975.

them with the medium-size and large power stations built by the State, the supply of electricity to the countryside should be gradually increased."

It was not until the beginning of the Leap Forward campaign in August 1958 that a plan for first-stage rural electrification was adopted at a national conference held in Tientsin. Under this plan a mass campaign for rural electrification was to be launched in five selected hsien and 100 selected APCs in each province or autonomous region during the winter and spring of 1958-59, and the first-stage rural electrification was to be completed during 1959-61.[53]

The conference also recommended that by the end of 1961, the total capacity of all electric power plants in the rural areas should reach 15-18 million kilowatts, producing 37.5-45 billion kilowatt-hours of electric power, so that the average annual consumption of electricity per person could reach 50-80 kilowatt-hours.[54]

The Tientsin conference also recommended that, as a means of popularizing rural electrification, emphasis be given to the construction of small hydroelectric plants, each with a generating capacity of not more than 50 kilowatts. It was believed that such plants would not require specialized technique, equipment, or critical materials.[55]

China's prospective water power resources provided another principal reason for emphasizing hydroelectric plants; these resources were first estimated by the National Resources Commission of the Nationalist government at 137 million kilowatts. The Communist government revised this figure several times: 140 million kilowatts in 1950, 150 million kilowatts in 1953, 540 million kilowatts in 1955, and 580 million kilowatts in 1958. However, these reassessments were made without considering such unfavorable conditions as the uneven distribution of precipitation in the PRC, which causes the waste of water in the high-water season and the decrease of power generation in the low-water season; the heavy investments needed for the construction of reservoirs, dams, generators, and transmission lines; and the difficulty of simultaneously maintaining supplies of water for irrigation and for power generation, especially in the north and northwest regions, which are subject to periodic droughts.[56]

Table 17.3 indicates that although the number of hydroelectric plants in rural China and their total capacity increased steadily from 1949 to 1958, the average capacity per plant declined as a result of the construction of small plants. At the end of 1958, there were 4,878 plants in the country's rural areas with a total capacity of 151,826 kilowatts, an average capacity of 31 kilowatts. Of these plants 4,334 with a total capacity of 131,502 kilowatts, an average capacity of 30 kilowatts, were built in 1958. Like the numerous native fertilizer plants and primitive blast furnaces for steel production constructed during the Leap Forward campaign, many of these small hydroelectric plants were ineffective. Some were constructed with only one water turbine, mostly made of wood, and had a generating capacity as small as three or four kilowatts.

Despite the Tientsin conference the development of rural electrification continued to be slow. In the first half of 1959, the total capacity of rural hydroelectric plants was increased by only slightly more than 30,000 kilowatts. The shortage of technical personnel and equipment was reported to be the major cause of the slow progress.[57] Probably the ineffectiveness of the small plants constructed in 1958 also hindered progress in this area. The development was further handicapped by the high cost of construction. Moreover, there was a lack of overall planning and of the accurate data needed for planning; surveys of hydraulic resources have been conducted only in some isolated regions.

As droughts spread to many provinces and autonomous regions in 1959, the operations of most of the hydroelectric power stations were badly affected. Losing all hope for the implementation of preliminary rural electrification planned for 1959-61, the Communist government had to adopt a new plan that called for "the realization of rural electrification to a considerable degree in a period of about ten years beginning in 1959."[58]

In addition to the small plants, a number of larger hydroelectric plants have been constructed in connection with projects for multiple purposes—flood control, and electric and water supply for industrial, domestic, and agricultural uses. In general these plants are considerably more efficient and economical than the small hydroelectric plants.[59] One of them is the Hsin-an River Hydroelectric Station, near Chienteh hsien in Chekiang Province, designed and constructed with Soviet assistance. Construction began in 1957 and the station went into operation in April 1960, when its first generator unit was installed. Since then the station has been supplying electric power to Shanghai, Nanking, Hangchou, and other cities and to pumping stations irrigating vast stretches of farmland in the Yangtze delta. It also serves to safeguard 300,000 mou of land along the lower reaches of the Hsin-an River from the menace of flood. The station was designed to have nine hydroturbine generator units with a total capacity of 652,500 kilowatts. The fourth generator, with a capacity of 72,500 kilowatts, was installed in 1963.[60]

The Liuchiahsia Hydroelectric Station in Yungching hsien, Kansu Province, China's biggest hydroelectric station, was completed at the end of 1974. Located on the upper reaches of the Yellow River on China's northwest plateau, it provides power to the provinces of Shensi, Kansu, and Chinghai. Built mainly to generate electricity from the Yellow River, it also regulates the river flow to prevent floods, reduces the menace of ice floes, supplies water to irrigate farmland, and makes possible the raising of fish and aquatic products. Astride the narrowest point of a gorge among steep mountains, the station has a concrete dam measuring 147 meters from top to bottom and is the highest in the country. One of the five huge water-turbine generating sets, all made in China, is the country's first 300,000-kilowatt set with an inner water-cooled stator and rotor. Total generating capacity is 1.225 million kilowatts, generating 5.7 billion kilowatt-hours of electricity a year.[61]

Early in 1975 China completed its first 330-kilovolt extra-high-tension power transmission line on the northwest plateau, which carried a powerful electric current from the country's biggest hydroelectric power station at Liuchiahsia on the upper reaches of the Yellow River to the rising industrial city of Tienshui in Kansu Province and the Kuanchung plain in neighboring Shensi Province. It crosses 14 hsien and operates through two major substations along the route. The new power transmission and transformation project set national records by its total length of 534 kilometers, its voltage of 330 kilovolts, and its transmission capacity of 420,000 kilowatts. The completion of the line helps to minimize the need for thermal electric power plants in the city of Tienshui and across the Kuanchung plain and to save over a million tons of coal per year. The line started to transmit power in June 1972. More than two years of practice and testing proved the soundness of the equipment and its installation and the good engineering quality of the project.[62]

Only a small number of thermal electric power plants have been constructed in the rural areas. Their total number at the end of 1958 was 2,112. One of the major problems here is that most of the equipment and tools used are more complicated and must be imported, whereas those for hydroelectric plants can be partly manufactured in China. Another difficulty is the shortage of coal, caused by the great demand in other industries and by the inadequacy of transportation facilities.[63]

In addition to the hydroelectric and thermal plants, another major source of electric supply in rural China is provided by the transformer substations of large electric power stations in urban and suburban areas. However, power transmission lines still are poorly developed, partly because of the shortage of metals and cement. Before the Leap Forward period, electric power networds of 35-, 10-6.6-, and 3.3-kilovolt transmission lines were the most common; but even these were rather scarce.[64] By the end of 1965, 126,000 kilometers of power transmission lines had been installed, as compared with 18,500 kilometers in 1957.[65] However, most of these power transmission lines are no more than local hookups serving small areas, rather than integrated hydrothermal power transmission systems.

One of the principal networks of high-tension transmission lines was installed in 1966 on the vast plain around Lake Poyang, an important area for producing rice, cotton, and oil-bearing crops in east China. The grid links the hydroelectric power stations in the area and the thermal power plants of nearby cities and towns. In a district of Nanchang hsien, 75.5 kilometers of high-tension transmission lines and 1,857 kilowatts of electric motors were installed to provide 6,600 hectares of farmland with electric pumping. The project was financed mainly by the eight communes of the district and undertaken by locally trained electricians, mechanics, and other technicians with state aid in the form of equipment, building materials, and technical personnel. Local carpenters and masons were enlisted for constructed work, and stones and rocks were substituted for cement to cut costs.[66]

For several years prior to the Cultural Revolution, the general policy of rural electrification was reliance on transmission lines as the primary source of electrical supply, supplemented by small hydroelectric stations.[67] The emphasis on transmission lines was due not only to the severe droughts in many parts of the country, which affected the operations of hydroelectric power stations, but also to the industrial recession in the early 1960s, which resulted in underutilization of the expanded generating capacity in the urban and suburban areas. As a part of the new policy of priority for agriculture, a program was set forth for the installation of transmission lines from the industrial and mining centers with surplus generating capacity to the surrounding rural areas.[68]

Subsequently the policy was reversed, mainly because of the contradictions between the demands of factories and mines and the needs for power supply of agriculture users. In accordance with Mao Tse-tung's instructions on self-reliance and along the same lines as the development of small chemical fertilizer plants and small farm machinery plants, increasing numbers of small hydroelectric stations again have been constructed or remodeled. These stations use largely local manpower and financial resources, with very little investment by the central authorities. This seems to indicate that industrial production was affected by the establishment of electric irrigation and drainage networks at the expense of urban power grids.

During and since the Cultural Revolution various localities have carried out overall planning and have built hydroelectric power stations on a large scale. This is in line with the principles that attention should be paid to the development of both thermal and hydroelectric stations and that more hydroelectric power stations should be built in places where hydraulic power resources are abundant. In the high mountainous areas in the upper reaches of the Yellow River, several large hydroelectric power stations have been completed with a total power-generating capacity of 1.3 million kilowatts, providing electric power to several hundred square kilometers. Another hydroelectric power station of an experimental nature is under construction in the silty middle reaches of the Yellow River. In the Yangtze River basin more than 20 large and medium-size hydroelectric power stations with a total generating capacity of 2.9 million kilowatts have been built along the Min River of the upper reaches of the Yangtze

River; along the Hsiang, Tsu, Yuan, Li, Han, and Kan rivers of the middle reaches of the Yangtze River; and along the Ching-i River of the lower reaches of the Yangtze River. Another group of hydroelectric power stations has been built in recent years along the Hun River of Liaoning, the Hsiao River of Hunan, the Hsiu River of Kiangsi, the Kutien River of Fukien, the Weng River of Kwangtung, and the Jung River of Kwangsi.[69]

In addition to the large and medium-size hydroelectric power stations, some 50,000 small ones, each with a capacity of 500 kilowatts, have been built in the countryside by various localities. Commune members in mountain areas use mountain creeks, small reservoirs, hillside ponds, and other water resources to generate electric power. Those on the plains dam rivers to raise the water head for generating power. Communes in coastal areas have built power stations that use tidal fluctuations as the motive force. Most of the small generating units are products of county- and commune-run workshops. Communes and production brigades generally train their own power station operators, while the state gives communes necessary assistance in investment and equipment. Technicians and workers from cities help with the designing of power stations and training of electricians.[70]

A mass movement for the construction of small hydroelectric stations in the rural areas took place throughout the country, following an on-the-spot meeting at Yungchun hsien, Fukien Province, convened in October 1969 by the Ministry of Water Conservation and Electric Power. Incomplete statistics indicated that the total power-generating capacity of medium-size and small hydroelectric stations built during the first half of 1970 was twice the total power-generating capacity of hydroelectric stations of all sizes built in the previous nearly 20 years. These stations were erected and financed mainly by the communes and production brigades, using indigenous methods and local materials.[71]

Rural consumption of electricity in the PRC in 1965 was reported to be 25 times that in 1957, increasing from 100 million to 2.5 billion kilowatt-hours. It rose sharply to 10.75 billion kilowatt-hours in 1973, 4.3 times that of 1965.[72] In 1965 electricity was available in most of the villages in about 1,300 of China's 2,126 hsien and benefited some 80 million farmers. The rapid increase was attributed to "the collective effort of the communes, energetically supported by the state"; in 1964 state investment in rural electrification was reported to be more than half the total funds appropriated for China's power industry as a whole.[73] However, it has been estimated that on the basis of five kilowatt-hours for each mou of cultivated land, China would need at least 8 billion kilowatt-hours each year for its 1.6 billion mou of cultivated land.[74]

Electric power is used in rural China primarily for the processing of agricultural products, especially rice husking and flour grinding, for the preparation of fertilizers and animal feed, and for illumination. The use of electricity for irrigation and drainage is increasing but still is not widespread. At the end of 1961, the total land area irrigated by electricity amounted to approximately 10 million mou (1.67 million acres), about 1 percent of the reported total area irrigated by all means. Most of the electrically irrigated farms were located in areas producing rice, wheat, cotton, or vegetables.[75] In 1965 electricity was used for irrigation and drainage on 84 million mou of farmland, about 16 percent of the reported total area irrigated by all means.[76]

Electric power also has been used on an experimental basis in the cultivation of land with a new type of plow operated by rope traction. It is claimed that, compared with tractors, the "electric plows" have the advantages of lower cost of manufacturing, less steel required, higher efficiency of operation (especially on paddy fields), and the possibility of operation by human and animal power, by wind, and by electricity.[77] But thus far no definite success has been revealed, although the use of the "electric plow" was expanded in Chekiang and a few other provinces the late 1960s.[78]

The use of tide, wind, marsh gas, and other sources of natural energy began to be popularized in July 1958. Special attention has been given to the transformation of tidal energy into electricity for agricultural uses. It is maintained that with a total coastline of more than 11,000 kilometers (6,830 miles), China has the possibility of generating some 34 million kilowatts of tidal energy. As an experiment several small tidal electric power stations have been built along the seacoast with capacities ranging from 15 to 250 kilowatts. Preliminary results indicate that although the construction of tidal electric stations is less expensive than that of ordinary hydroelectric plants, tidal energy is not dependable because of the wide fluctuations of the tide.[79]

Thus rural electrification in the PRC, which started virtually from scratch, progressed slowly but steadily in the 1950s and 1960s. There is little doubt that electricity, as a less expensive form of power, will become increasingly important in both agricultural production and household uses in the rural areas. In view of the country's rich hydraulic resources and the constant shortage of coal, metals, and cement, the principal sources of electricity in rural China probably will continue to be small hydroelectric plants, supplemented by transformer substations of large electric power stations in urban and suburban areas. Therefore, more rapid development of China's rural electrification will depend largely on the improvement of small hydroelectric plants and the expansion of a number of large projects to serve the principal crop-producing areas, as well as on the development of an integrated hydrothermal power transmission system on a nationwide scale.

THE PROSPECTS

In view of the slow progress of the tool innovation program, remedial measures were taken following the crop failure of 1959-61. The First National Conference for Scientific Discussion on Agricultural Machinery was held in Peking for three weeks in June and July 1962. Participants included Vice-Premier Teng Tzu-hui; Han Kuang, vice-chairman of the State Science and Technology Commission; Ch'en Cheng-jen and Chang Feng-shih, minister and vice-minister of Agricultural Machinery, respectively; and other high government officials. This conference was followed by the establishment of the Chinese Academy of Agriculture Mechanization in Peking on July 16, 1962.[80] Research institutes of agricultural mechanization have also been set up in various provinces and autonomous regions.[81]

The problem of agricultural mechanization was again thoroughly discussed at the Eighth Plenary Session of the Tenth Central Committee of the CCP, held in September 1962. Participants in the meeting recognized that the "arduous and complicated" task of mechanization of agriculture would require a period of 20 to 25 years, as originally planned, rather than ten years, as revised during the Leap Forward period, and that semimechanized and improved farm tools would remain in a very important position for a long time to come.[82] Subsequently the policy of "simultaneous development of mechanization and semimechanization, with the main stress on the latter," has been implemented. In the PRC "semimechanization" generally refers to the use of machines and implements with improved working mechanism that still must be operated mainly by human or animal power. Semimechanization includes many types of farm implements and tools of 20 horsepower or less for sowing, cultivation, plant protection, irrigation and drainage, harvesting and threshing, rural transportation, and processing of agricultural products.[83] Most of these semimechanized implements and tools are manufactured locally: in 1966 the total output value of local agricultural machinery plants accounted for more than two-thirds of that of the agricultural machine-building industry as a whole.[84]

In 1965 there were more than 25,000 handicraft factories and workshops employing a total of 810,000 persons and producing farm machines and implements to suit local needs, according to Teng Chieh, vice-minister of the Second Ministry of Light Industry.[85] Chia An-lan, an alternate member of the National Committee of the

All-China Federation of Handicraft Cooperatives, revealed that in the first nine months of 1965 a total of more than 1.3 million semimechanized farm machines and tools were made by handicraft works and cooperatives in 19 provinces and municipalities. He pointed out many advantages of semimechanized farm implements and tools over large and more elaborate farm machines, including adaptability to different types of land, transportability due to small size, simplicity of construction, and lower cost.[86]

The policy of agricultural mechanization was clearly reflected at the National Agricultural Exhibition held at Peking in 1966. The approximately 1,300 farm machines and implements on display were designed with local conditions taken into account; they were of all sizes, but the small ones formed the bulk.[87]

Meanwhile, in order to improve maintenance and repair facilities, a number of agricultural machinery accessories plants were constructed or reconstructed after 1960, including the Shihchiachuang (Hopei) Tractor Accessories Plant, the Wusih Oil Pipes and Oil Pumps Plant, the Nanch'ang Tractor Accessories Plant, the Pengpu (Anhwei) Tractor Accessories Plant, the K'aifeng Tractor and Electrical Machinery and Tools Factory, and the Luk'ou (Hunan) Internal Combustion Machine Accessories Factory. A large portion of capital construction investment in the agricultural machinery industry has been allotted for accessories plants.[88] By the end of 1965, agricultural machinery and tractor stations were reported to have been set up in seven of every ten hsien in the PRC.[89] But in 1966 the supply of spare parts for farm machinery remained inadequate in many areas.[90]

In 1975, 96 percent of all hsien had farm machinery repair plants, and most communes and production brigades had their own farm machinery repair centers or workships. Many provinces, municipalities, and autonomous regions had set up small cement, iron, and steel plants as well as coal pits to provide materials and services for producing and repairing farm machines.[91]

Although the revised plan for agricultural mechanization is much less ambitious than the plan mapped out during the Leap Forward campaign, whether or not it can be completed according to the schedule—that is, by the end of 1980—remains to be seen. Two basic questions arise: (1) how many agricultural machines and implements can be produced and (2) how much mechanization agriculture will be able to absorb profitably. The shortage of technical personnel, and of steel and other raw materials, and the somewhat improved but still inefficient management under the communes undoubtedly will continue to give the Chinese Communists difficulties in regard to the production and maintenance of agricultural machinery and tools. Meanwhile, it is generally believed that economic and social conditions in the PRC probably will remain immature for a nationwide agricultural mechanization for some time to come. As the People's Daily pointed out, "On the one hand, the level of our industrial production still cannot supply a sufficient amount of farm machinery for agricultural production; and on the other hand, it is necessary to raise the cultural and technical level of the peasants if farm machinery is to be put into extensive use in the rural areas."[92]

To what extent and in what manner Chinese agriculture should be mechanized is a matter that needs careful consideration. Simple small-scale mechanization of the type being developed in Japan, for example, may prove more economical than the employment of human labor and the maintenance of draft animals. Mechanization also may contribute to higher yields by permitting more timely operations when the soil is in the best condition for cultivation or the weather is favorable for harvesting. Furthermore, mechanization may free for other crops land that was formerly needed to provide livestock feed. For some operations the greater efficiency of mechanization may outweigh other considerations. For example, deep plowing, which can help improve soil fertility and increase yields under certain conditions, is difficult without heavy equipment. Many modern pesticides cannot be applied effectively without improved types of sprayers and dusters, and most modern mechanized planters and fertilizer appliers do a much better job of seeding and fertilizing than traditional manual methods do.

On the other hand, sufficient attention should be paid to evolving types of farm equipment specifically adapted to the country's conditions and needs. This includes the use of garden-type tractors, small harvesting and threshing machines, and motorized tricycles in rice-producing in the south and densely populated areas in the north, in order to meet local conditions of small field plots and highly intensified cultivation. Research on mechanization is required to provide guidance as to the most appropriate forms of mechanization and thus help to avoid repetition of the costly mistakes that the PRC has made.

In view of the PRC's limited industrial capacity and other difficulties, the present policy of "simultaneous development of mechanization and semimechanization, with the main stress on the latter," for 20 or 25 years (1962-82 or 1962-87) seems both more feasible and more profitable than the overambitious plan mapped out during the Leap Forward campaign, which called for complete mechanization within the ten years 1959-68.

NOTES

1. Government of the Republic of China, Chungking, "Program and Estimated Requirements for Relief and Rehabilitation in China,"

presented to the United Nations Relief and Rehabilitation Administration (Washington, D. C., 1944), Annex F, "Agriculture," pp. 75-81. (Mimeographed.)

2. Li Ching-yu, "Gradual Realization of Agricultural Mechanization Through Tool Innovation," Nung-yeh Chi-hsieh [Agriculture Machinery], Peking, No. 18, 1959, pp. 1-5.

3. Ni Ts'ai-wang, "How to Use Double-Wheel/Double-Share Plows and Double-Wheel/Single-Share Plows?" Sheng-wu-hsueh T'ung-pao [Biological Bulletin], Peking, September 1956, pp. 65-68.

4. "Why the Demand and Production of Double-Wheel/Double-Share Plows Have Stopped?" editorial, Chi-hua Ching-chi [Planned Economy], Peking, No. 9, 1956, pp. 1-4.

5. "Raise the Effectiveness of the Double-Wheel/Double-Share Plows," editorial, JMJP, April 16, 1958, p. 1; Hsin-hua Pan-yueh-k'an [New China Semimonthly], Peking, No. 9, 1958, pp. 86-88.

6. "Ten Years of Great Accomplishment and Experience in Farm Machinery Work in China," Nung-yeh Chi-hsieh, No. 18, 1959, pp. 10-14.

7. Nung-yeh K'o-hsueh T'ung-hsun [Agricultural Science Bulletin], Peking, No. 2, 1958, pp. 62-63.

8. China Reconstructs, Peking, No. 5, 1957, p. 9.

9. Peking Review, No. 39, 1973, pp. 22-23; NCNA-English, Peking, June 1, 1974.

10. NCNA-English, Peking, July 26, 1973.

11. Liu Jui-lung, "The Revolutionary Significance of Rice Transplanters," K'o-hsueh Ta-chung [Popular Science], Peking, No. 4, 1960, pp. 133-35; NCNA-English, Peking, April 27, 1969.

12. Li Ching-yu, op. cit.

13. "Grasp Firmly the Repair and Manufacturing of Agricultural Machinery and Implements," editorial, JMJP, November 15, 1960, p. 1.

14. JMJP, December 26, 1960, p. 3.

15. Ibid., November 18, 1960, p. 3.

16. Ibid., December 2, 1960, p. 2.

17. Ibid., November 16, 1960, p. 1; November 18, 1960, p. 3.

18. Ibid., December 29, 1960, p. 2.

19. "On the Question of Agricultural Cooperativization," Jen-min Shou-ts'e [People's Handbook] (Peking, 1956), pp. 80-86.

20. State Statistical Bureau, Ten Great Years (English ed.; Peking, 1960), p. 135.

21. Hsin-hua Pan-yueh-k'an, No. 17, 1959, p. 24.

22. Ch'en Cheng-jen, "Speed up Technical Transformation of Agriculture," Hung Ch'i [Red Flag], Peking, No. 4, 1960, pp. 4-10.

23. JMJP, April 11, 1960, p. 3; October 28, 1960, p. 4.

24. "Actively and Systematically Bring About the Technical Transformation of Our Country's Agriculture," editorial, JMJP, November 9, 1962, pp. 1-2.

25. NCNA-English, Peking, September 30, 1972; "New Leap in China's National Economy," Peking Review, No. 2, 1972, pp. 7-9.

26. JMJP, October 27, 1960, p. 1.

27. Writing Group of the Heilungkiang Provincial Revolutionary Committee, "Struggle for the Fulfillment of the Great Task of Agricultural Mechanization at an Accelerated Pace," NCNA Domestic Service, Peking, August 7, 1970.

28. "Actively and Systematically Bring About the Technical Transformation of Our Country's Agriculture," op. cit.

29. State Statistical Bureau, op. cit., p. 98.

30. NCNA-English, Peking, January 31, 1966.

31. China News Service, Canton, news release No. 4146, August 9, 1965.

32. JMJP, April 28, 1962, p. 1.

33. NCNA, Peking, February 5, 1966.

34. Teng Chieh, "Bring the Role of Handicrafts into Full Play and Serve Agricultural Production Better," JMJP, January 4, 1966, p. 5.

35. Hsu Pao-yuan, "How Are the More Than 700 Types of Rolled Steel Connected with Agriculture?" Shih-shih Shou-ts'e [Current Events Handbook], Peking, No. 17, 1962,

36. JMJP, January 7, 1962, p. 1.

37. Chao Hsueh, "Current Problems in Agricultural Mechanization," Ta Kung Pao, Peking, May 15, 1961, p. 3.

38. Hsiang Nan, "Certain Problems of Agricultural Mechanization," JMJP, December 22, 1962, p. 5.

39. JMJP, December 18, 1962, p. 2.

40. Ibid., December 8, 1962, p. 2.

41. Teng Chieh, op. cit.

42. "It Is Fine to Have Tractors Back to Chairman Mao's Revolutionary Line—A Report of Investigation of the Re-collectivization of the Operation of Tractors in Lankao Hsien," Nung-yeh Chi-hsieh Chi-shu [Technology of Agricultural Machinery], Peking, No. 10, 1968, pp. 4-7.

43. JMJP, November 5, 1960, p. 7.

44. Chang Feng-shih, "Produce Suitable Agricultural Machinery to Support Agriculture," Kung-jen Jih-pao [Daily Worker], Peking, January 18, 1962, p. 1.

45. Yang I-ch'eng, "Positively Develop Draft Animals and Accelerate Agricultural Production," JMJP, May 9, 1961, p. 7.

46. JMJP, July 28, 1960, p. 2; March 27, 1961, 1. 2.

47. "Protect Safe Wintering of Draft Animals," editorial, JMJP, December 23, 1960, p. 1.

48. JMJP, April 12, 1961, p. 3.

49. Hsu Chi, "Accomplishments of Animal Husbandry in the Past Decade," Chung-kuo Hsu-mu-hsueh Tsa-chih [China Animal Husbandry Magazine], Peking, No. 10, 1959, pp. 289-94.

50. JMJP, November 17, 1962, p. 2; "Properly Manage Livestock Breeding and Veterinary Stations," editorial, JMJP, February 17, 1963, p. 1.

51. G. Meisinger, "Training of Veterinarians in the Chinese People's Republic," Monatshefte fuer Veterinaer-Medizin [Monthly Bulletin of Veterinary Medicine], Jena, No. 1, 1966, translation by Joint Publications Research Service, JPRS 34,331 (Washington, D. C., March 1, 1966), pp. 7-11.

52. "Celebrating the Tenth Birthday of Our Great Nation," editorial, Chung-kuo Shou-i-hsueh Tsa-chih [Chinese Veterinary Science Magazine], No. 9, 1959, pp. 253, 261.

53. Hsin-hua Pan-yueh-k'an, No. 17, 1958, p. 97.

54. "A Bugle Is Blown for the March Toward Rural Electrification," editorial, Shui-li Fa-tien [Hydroelectrics], Peking, No. 17, 1958, pp. 3-5.

55. Ho Chi-feng, "Summary of Report to the National Conference on Rural Electrification," Shui-li Fa-tien, No. 17, 1958, pp. 6-7.

56. Robert Carin, Power Industry in Communist China (Hong Kong: Union Research Institute, 1969), pp. 36-40.

57. Chang Chung-wei et al., "Communization Is Beneficial to Rural Electrification," JMJP, January 26, 1960, p. 7.

58. Liu Lan-p'o, "The Road to the Realization of Rural Electrification," Hung Ch'i, No. 13, 1960, pp. 8-15.

59. Shui-li Fa-tien, No. 12, 1958, p. 54.

60. Hsu Hsia-shih, "The Hsin-an River Hydroelectric Station," Ti-li Chih-shih [Geographical Knowledge], Peking, No. 6, 1957, pp. 257-59; "1963 Successes of China's Economy," Peking Review, No. 40, 1963, pp. 17-21; JMJP, September 23, 1963, p. 2; "15 Years of Hsinan Kiang Hydro-power Station," Peking Review, No. 31, August 1, 1975, p. 22.

61. "China's Biggest Hydro-Power Station," Peking Review, No. 7, 1975, pp. 11-12, 21.

62. NCNA-English, Peking, February 6, 1975.

63. B. Bannikov, "Rural Electrification in the Chinese People's Republic," Tekhnika v sek'skom khoziaistve, [Technology and Agriculture], Moscow, No. 2, 1960, translation by Joint Publications Research Service, JPRS 4529 (Washington, D. C., April 12, 1961), of an article in pp. 88-92.

64. Ibid.

65. "The Practice of Operating the Electricity Service in Rural Areas on a Part-Work and Part-Farming Basis," editorial, *JMJP*, September 7, 1965, p. 1.

66. NCNA-English, Nanchang, June 4, 1966.

67. Chung Hsin, "New Development of Rural Electrification in China," press release, China News Service, Canton, February 9, 1965.

68. Tso Hu, "Several Problems of Agricultural Electrification," *Ching-chi Yen-chiu* [Economic Research], Peking, No. 3, 1963, pp. 9-16.

69. NCNA Domestic Service in Chinese, Peking, November 1, 1973.

70. NCNA-English, Peking, January 29, 1973.

71. *JMJP*, July 18, 1970, p. 1.

72. NCNA-English, Peking, December 25, 1973.

73. *JMJP*, December 25, 1965, p. 1; NCNA's special report on the occasion of the sixteenth anniversary of the founding of the PRC, Peking, September 16, 1965.

74. Liu Jih-hsin, "A Discussion of Several Problems Concerning the Modernization of Our Country's Agriculture," *JMJP*, June 20, 1963, p. 5.

75. *JMJP*, May 11, 1961, p. 3; December 23, 1961, p. 1.

76. NCNA, Peking, September 16, 1965; *Peking Review*, No. 41, October 8, 1965, p. 29.

77. "Energetically Popularize Rope Traction Machines," editorial, *Nung-yeh Chi-hsieh*, No. 7, 1958, p. 1; "A Rapid Way Toward the Mechanization and Electrification of Cultivation," editorial, *JMJP*, September 12, 1958, p. 1; *Nung-yeh K'o-hsueh T'ung-hsun*, No. 10, 1958, pp. 518-20.

78. *Ching-chi Tao-pao* [Economic Reporter], Hong Kong, March 6, 1967, p. 16.

79. Chang Han-ying, "Report to the National Conference on the Transformation of Tidal Energy into Electricity, Shanghai, October 1958," *Shui-li Shui-tien Chien-she* [Water Conservation and Hydroelectric Construction], Peking, No. 2, 1958, pp. 3-5.

80. *Nung-yeh Chi-hsieh Chi-shu*, No. 5, 1962, pp. 2-3.

81. NCNA-English, Peking, February 7, 1966.

82. "Actively and Systematically Bring About the Technical Transformation of Our Country's Agriculture," editorial, *JMJP*, November 9, 1962, pp. 1-2.

83. "What Are Semimechanized Farm Machines and Tools?" *JMJP*, November 9, 1964, p. 2.

84. NCNA, Peking, October 17, 1966.

85. Teng Chieh, op. cit.

86. Chia An-lan, "Produce and Popularize Semimechanized Farm Tools on the Basis of Reality," JMJP, January 4, 1966, p. 5.

87. Liu Pang, "Paving a Way to Agricultural Mechanization in China—A Visit to the Agricultural Mechanization Hall of the National Agricultural Exhibition," JMJP, April 13, 1966, p. 2.

88. JMJP, August 1, 1963, p. 2; Kuang-ming Jih-pao [Kuang-ming Daily], Peking, September 2, 1963, p. 4.

89. NCNA-English, Peking, December 28, 1965.

90. "Do Not Underestimate the Problem of Parts," JMJP, April 12, 1966, p. 2.

91. NCNA-English, Peking, November 13, 1975.

92. "The Spirit of Self-Reliance Must Be Upheld in the Movement of Farm Implement Innovation," editorial, JMJP, October 15, 1964, p. 1.

CHAPTER

18

FIELD MANAGEMENT

Kuan, or field management, is the last and also the least discussed of the eight major aspects of Mao Tse-tung's "agricultural charter." It generally refers to the meticulous management of farmland from the time of sowing or transplanting to the time of harvest, including the coordination and timely application of horticultural techniques used by Chinese farmers for many generations to promote factors and conditions favorable to agricultural production. The purpose of field management, according to the Research Institute of Crop Breeding and Cultivation, is "to adjust the physiological activities of crops during various stages of their growth so that a maximum yield of the crops may be obtained."[1]

Provisions on field management in the NPAD were very general and not very clear. Article 11 stated the following:

> All agricultural producers' cooperatives should undertake intensive and meticulous farming and should in a rational way improve their methods of cultivation, fertilization, irrigation, crop rotation, inter-cropping, multiple-cropping, and close planting. Efforts should also be made to ensure that field work, such as sowing, weeding, and thinning, will be done in right time, that field management will be strengthened, and that harvesting as well as threshing will be carried out promptly with care in order to ensure high yields and bumper harvests.

More specifically, field management requires concern with the following matters during the growing period of crops: (1) conditions essential for the germination of seeds, including the supply of moisture and air and the elimination of soil crust; (2) suitable moisture, air, nutrients, and soil temperature; (3) weed control; (4) elimination or lessening of unfavorable meteorological conditions, such as frost, drought, and waterlogging; (5) intertillage, that is, cultivation of land between the rows of crops; (6) thinning of seedlings to space them reasonably evenly, especially for cotton, jute, corn, kaoliang, and millet; (7) ridging of ground to protect plants against cold, waterlogging, ineffective offshoots, growth of weeds, and insect pests and diseases; and (8) pinching off of leaves to control the growth of the upper part of plants and to concentrate nutrients on their lower part, especially for rapeseed, tobacco, cotton, and sweet potatoes and other vegetables.[2]

One of the most frequently quoted examples of field management is the "three-yellow and three-black" technique of controlling the growth of the single-crop late-ripening variety of paddy rice, devised by Ch'en Yung-k'ang (see Chapter 3). Ch'en, a farmer of Kiangsu Province, gained nationwide fame as a model peasant because he achieved one of the country's highest yields of paddy rice. According to his observations the color of shoots of the late rice crop changes three times from yellow to black (or dark green) and three times from black to yellow during the period of growth. The schedule of the changes in color helped Ch'en to determine when and how to use additional manure and to water or drain the fields in order to promote a high yield. In other words, the management of rice fields should be based on observations of the subtle changes of leaf color in order to obtain high yields.

Another example of good field management is the use of the shape of seedlings at different stages (pig ear, donkey ear, and horse ear) by a Honan peasant, Liu Ying-hsing, as the basis for fertilizing and watering the wheat crop. When the leaves of wheat seedlings are in the shape of pigs' ears, dark green in color, and hanging down, it is an indication of overgrowth and the fields should be watered sparsely. When the leaves are in the shape of donkeys' ears, bluish-green in color, and with tops slightly bent over, it is an indication of normal growth and the fields should be fertilized with additional nitrogen, phosphate, and potassium in proper proportion and should be watered. When the leaves are in the shape of horses' ears, light green in color, and standing straight, it is an indication of fertilizer deficiency and the fields should immediately be fertilized with large quantities of nitrogen and should be proportionately watered at the same time.[3]

In urging good field management of winter wheat in 1961, the _People's Daily_ stated:

> The promising growth of the wheat seedlings had caused some people to slacken efforts in field management in

> the belief that a bumper harvest is already certain. This is wrong. Good seedlings are only the foundation for increased output. Wheat requires a relatively long period to grow and mature. During the long winter and spring months, the wheat has to face frequent and varied natural calamities, and needs sufficient nutrients and tender care.
>
> Measures for field management of wheat depend on local conditions. To hasten the growth of wheat, give it water if it needs water and give it additional fertilizer if it needs fertilizer. On the other hand, land which is too fertile should be so controlled so as to retard the growth of wheat during the early stage. Dig drainage ditches if there is too much rain and water. All these tasks must be properly carried out during the growth of the wheat seedlings.
>
> Meanwhile, preparations should be made for late-stage wheatfield management next spring. The winter months are especially important for making preparations to fight wheat diseases and pests and for carrying out the installation and repair of irrigation projects. What is more important, the fertilizer needed by wheat during the coming spring when it begins to turn green and grow rapidly should be gathered in the winter, since fertilizer gathered then, after a period of fermentation, can be used immediately after the melting of the snow and ice in spring. If we miss the winter season and rely on spring for gathering fertilizer, there will not be enough time for the fertilizer to ferment.[4]

In the case of cotton, the People's Daily, in another editorial, stated the following: "Field management should be strengthened to insure the healthy growth of cotton seedlings. Tilling and weeding should be carried out expeditiously during the early period of cotton cultivation because of the quick evaporation and the fast growth of weeds around the cotton seedlings. Even a day's delay can mean the difference. The most suitable time to apply additional fertilizer to cotton fields is in June or around the first half of July, the early and intermediate periods of cotton growth. Control of insect pests in cotton fields should be carried out in a timely manner. Missing this opportunity might bring about a need for several times more manpower and material to do the job later and possibly without much effect."[5]

In acclaiming the achievements of field management, Vice-Premier T'an Chen-lin stated: "The people's communes are now everywhere making big efforts to cultivate large high-yielding tracts and to exercise meticulous care in farming, just as in gardening. The peasants say that the farms now look like vegetable gardens with deep plowing, close planting, good seed strains, proper manuring, proper irrigation, careful cultivation and intensive farming, and elimination of weeds, insect pests and plant diseases."[6]

Despite the increasing emphasis on the modernization of agriculture in recent years, the communes have been repeatedly asked to give 30 percent of their efforts to sowing and 70 percent to field management.

NOTES

1. Research Institute of Crop Breeding and Cultivation, Chinese Academy of Agricultural Science, "Field Management," Hsin-hua Pan-yueh-k'an [New China Semimonthly], Peking, No. 12, 1959, pp. 129-30.

2. Chu Chien-nung, T'u-ti Fei-li Ching-chi Yuan-li [Economic Principles of Land Fertility] (Shanghai, 1964), pp. 298-99, 328-33.

3. Ibid., pp. 298-99.

4. "Grasp the Important Link in Increasing Wheat Production," editorial, JMJP, December 27, 1961, p. 1.

5. "Momentous Juncture in the Raising of the Cotton Output," editorial, JMJP, June 10, 1963, p. 1.

6. T'an Chen-lin, "Strive for the Fulfillment, Ahead of Schedule, of the National Program for Agricultural Development," JMJP, April 9, 1960, p. 1.

CHAPTER

19

THE PROSPECTS FOR RAISING THE YIELD PER UNIT OF LAND

China is reaching the limit of the area that can be cultivated economically and is extending into very marginal land. The "uncultivated arable land" is mostly of poor quality and can be brought into production only after heavy investment. Reclamation of such land is often a costly and lengthy process. Therefore, the more effective means to improve agricultural production would be to increase the productivity of the land already cultivated. For this reason most measures prescribed in the NPAD were geared to raising the yield per unit of land.

Although many accomplishments have been claimed in all eight aspects of the "agricultural charter" that Mao Tse-tung considers the concentration areas of the program, the preceding chapters clearly indicate that none of these measures has progressed as successfully as expected. Moreover, in adopting new techniques and inputs, the Communist government has not paid sufficient attention to the principle of complementarity and the need to introduce new approaches in an integrated rather than piecemeal fashion. For example, outside the paddy rice area in China, only a limited portion of the lands for grain crops can at present make economical use of chemical fertilizers without irrigation. Also the use of new varieties of crops in many cases requires heavier application of fertilizers, adequate irrigation, timely control of plant diseases and insect pests, and a broad range of improved farming practices. The full impact of these measures can be obtained only when they are combined in a suitable "package."

As a whole, Mao Tse-tung's eight major aspects of the technical transformation of agriculture progressed at only a moderate pace during the First Five-Year Plan period, and progress was slowed by the Leap Forward campaign and the crop failures of 1959-61. The recovery since then has been slow but steady. It is likely that progress in the 1970s will accelerate somewhat as a result of the adoption of remedial measures in recent years and the improvement of general economic conditions.

CHAPTER

20

SUMMARY AND CONCLUSIONS

Since the early 1950s the Communist regime has made continuous efforts to transform China "from a backward agricultural country into an advanced socialist industrial country." Until recent years agriculture in the PRC had not been given as much attention as it should have received. In capital investment, agriculture was given much lower priority than heavy industry in the First Five-Year Plan and during the Great Leap Forward campaign. Labor, financial, and distribution systems under centralized management were inefficient, particularly during the early period of the communes. The program for improving agricultural techniques had progressed very slowly because of improper planning and execution and because of shortages of funds, personnel, and materials. It was not surprising that the First Five-Year Plan period was concluded with spectacular achievements in heavy industry, moderate success in light industry, but only slight improvement in agriculture.

After the 1959-61 crop failure many remedial measures were taken to increase agricultural production. As a result of the improvement, as well as a favorable weather and economic situation in general, harvests of most of the principal crops between 1962 and the Cultural Revolution were reported to have attained—or nearly attained—the level of 1957, the year of highest agricultural production prior to the Leap Forward campaign. The Communist government has also claimed that for the fourteen consecutive years since 1962, China had had bumper harvests. In a frequently quoted statement made to Edgar Snow in March 1971, Chou En-lai said that China's total grain output in 1970 was more than 240 million tons and that the country had accumulated state grain reserves of about 40 million tons.[1] A New China News Agency dispatch of December 31, 1971, indicated that despite serious natural calamities China's grain output in 1971 reached 246 million tons. The 1970 and 1971 production figures were thus 55 million tons and 61 million tons more, or 30 percent and 33 percent higher, respectively, than the 185 million tons for 1957. Grain production figures for 1972 and 1973 have not been announced, but the output in 1974 was reported to have reached nearly 275 million tons. Although interpretations and evaluations of these figures have varied considerably among different sources, agricultural production in the PRC is generally believed to have grown steadily since the mid-1960s.

In the previous chapters a perspective on the agricultural development during the first quarter-century of the PRC has been presented by reviewing two interrelated and complementary types of transformation of agriculture considered by the Communist authorities as essential to increase China's agricultural production. On the basis of such a perspective, an effort has been made to determine, among other things, the extent to which institutional changes and technical innovation have helped agricultural productivity in the PRC and the prospects for the next few decades.

Following the completion of land redistribution in 1952, the masses of peasants were led along the road to cooperativization and communalization, step by step, each step leading farther away from traditional land tenure and management. Whether or not the successive institutional changes have improved agricultural production is difficult to gauge because of lack of data. Moreover, each of the transition periods was so brief that there was no sound basis for a comparison with previous records. It is clear, however, that the farm crisis in 1959-61 was partly caused by the improper management of the newly established communes and was aggravated by natural calamities and the withdrawal of Soviet assistance.

As a result of the structural modification of the commune system, beginning in the latter part of 1959, the production team has become the basic unit, both with respect to the collective ownership of means of production and the unified distribution of income and products; and the functions of the commune and the production brigade have been drastically reduced. The shifting of functions has permitted the peasants to suffer less interference and to receive a larger share of income and products. It has stimulated their incentive to work and is believed to have helped agricultural production in general

Whether or not the relaxation of centralized direction will continue to prevail remains very uncertain. The present p'i-lin p'i-k'ung campaign (criticism of Lin Piao and Confucius), which began in early 1974, has not changed the moderate agrarian policy adopted in 1971, but within that framework stricter and more radical measures have been introduced. Since the early 1970s the emphasis has been on developing collective economy and on restraining the independent activities both of the production teams and of the individual cadres and peasants. For example, Article 7 of the new constitution of the People's Republic of China, adopted on January 17, 1975, continued to provide the three-level ownership of the commune with the production team as the basic accounting unit. It also sanctioned private plots for personal needs, limited household sideline production, and, in pastoral areas, a small number of livestock, "provided that the development and absolute predominance of the collective economy of the people's commune are ensured." Despite these provisions, in February 1975, only a month after the adoption of the new constitution, the emphasis on the proletarian dictatorship was clearly indicated in a nationwide campaign. In several articles printed in the February 1975 issue of Hung Ch'i, the authors sternly warned that the little freedoms of the peasants—the private plots and the domestic side-occupation, sanctioned by the new constitution—must not be allowed to get out of hand. On February 10, 1975, the People's Daily published an article in which the writer said that the peasants' private ownership must be restricted, not expanded: "The traditional concept of private ownership must be smashed to the foundations" and "this garbage must be swept away."[2] As of the spring of 1976, Peking was still carefully instituting a new system of material incentive to encourage agricultural production, but at the same time it was desperately trying to convince the peasants that this system was not reverting to the system sponsored by "swindlers" like Liu Shao-ch'i. The communes have been frequently reminded of "implementing properly" the rural economic policy adopted early in 1971.

Although the technical transformation of agriculture during the first quarter-century of the PRC was rather limited as compared with the vast backlog of traditional Chinese agriculture and had benefited overall production to only a small degree, it has become an increasingly important factor in determining the level of agricultural output. The technical advances of agriculture in the PRC have been handicapped by shortages of trained personnel, of funds for capital construction investment and the purchase of the means of production, and of certain necessary materials and equipment, including rolled steel, copper, cement, and lumber; and economic and social conditions in some areas were not conducive to modernization. The situation, however, has gradually improved in recent years.

One of the important measures taken since the 1959-61 crop failure in an effort to expedite the technical transformation of agriculture is the adjustment whereby heavy industry is made to serve agriculture. The government has repeatedly ordered that, above all, industry must satisfy the needs for technical innovation of agriculture. Data on industrial aid to agriculture are sketchy, but it is clear that the program involves many serious economic problems. First, it is an extra burden added to the already heavy tasks of industry. Steel, copper, cement, timber, and benzene, for example, are badly needed in many other fields as well as agriculture. In addition, there is the question of the extent to which industry can change its objectives and structure without suffering financial loss and reducing production efficiency. This leads to another question: Even with the readjustment of heavy industry, how much machinery, equipment, and materials for agriculture can be produced?

Another important measure taken in recent years to speed up the technical transformation of agriculture was the promotion of small-scale industries in the rural areas to supplement the industrial aid to agriculture program adopted in 1960. In order to make the communes and the production brigades as self-sufficient as possible in the supply of the materials and services needed for agricultural uses, the policy of simultaneous development of large and small industrial enterprises has been implemented. The latter includes the small factories producing cement, chemical fertilizers, iron, and steel, the small plants manufacturing and repairing farm machines and tools, small hydroelectric power stations, and small coal mines and pits. In 1973, nearly half of China's 2,000 and more hsien had set up small chemical fertilizer plants, which accounted for half of the country's synthetic ammonia; two-thirds of the hsien had small cement works, producing about half of the country's total; small farm machinery works and repair stations had been built in more than 90 percent of the hsien; every province, municipality or autonomous region had built at least one small iron and steel mill; and many hsien had opened small coal mines and hydroelectric power stations.[3] China cannot solely rely on large plants to meet agricultural requirements. These small plants can be built and put into operation in a relatively short time.

What is even more significant is the fact that in recent years, the PRC has appeared to be in a continuous process of learning, mainly by trial and error, the techniques and inputs suited to its own circumstances and stage of agricultural development. It would seem that, if such efforts continue, yields of grain and other principal crops could be raised considerably within the next few decades, both because of the country's present low technical levels and because of the availability of new production techniques and inputs from modern agricultural science. Yields in other countries with similar situations have on occasion increased substantially within a short time period.

But realization of agricultural progress will require large-scale investment of manpower, funds, and other resources, also organization or reorientation of a great number of activities and coordination of national, provincial, and local efforts. Raising a country's agricultural productivity is both a technical problem and a social and economic problem. It is also a problem of organization. The principal obstacle to agricultural development in many countries is not lack of technical knowledge but the problem of transmitting that knowledge to millions of farmers. Perhaps still more important is the creation of an environment in which farmers will have the incentive and also the means to put improved methods of farming into practice.

The People's Republic of China, inheriting a war-torn agriculture with depleted resources and insufficient personnel, has, within a short period of twenty five years, succeeded in transforming its disorganized and impoverished rural economy into production units and in evolving a social and economic system that allows a certain degree of increased consumption and economic incentives. In recent years, the government has also made considerable improvement in the development of farming techniques. It would appear that when all important social, economic, and technological factors have been taken into consideration—barring unforeseen developments—the present trend of accelerating increase in agricultural production will probably continue to prevail for the next few decades. In the light of rapid population growth, industrial expansion, and requirements for exports, however, even the continuation of the present trend of agricultural production would still leave the PRC an enormous task in the supply of farm products.

NOTES

1. Edgar Snow, "Talk with Chou En-lai: The Open Door," New Republic, March 27, 1971, pp. 20-23.
2. Liang Hsiao, "Proletarian Dictatorship Against Capitalist Trends Must Be Realized," JMJP, February 10, 1975, pp. 1, 3.
3. NCNA-English, Peking, September 18, 1973.

APPENDIX

THE NATIONAL PROGRAM FOR AGRICULTURAL DEVELOPMENT, 1956-67

Introduction

This is a programme of endeavour to bring about a rapid increase of the agricultural productive forces, so as to reinforce socialist industrialization in our country and raise the living standards of the peasants and the people as a whole during the period from the First Five-Year Plan to the Third.

Socialist industry is the leading force of our national economy. But the development of agriculture occupies a vital place in our socialist construction. Agriculture supplies industry with grain and raw materials. At the same time, the countryside, with its more than 500 million population, provides our industry with the biggest domestic market in the world. In this sense, without our agriculture there could be no industry in our country. It is utterly wrong to belittle the importance of agricultural work.

There are two ways to develop agriculture. One is the capitalist road whereby the fate of the peasants is in the hands of landlords, rich peasants and speculating merchants and a handful of people get rich while the vast majority are driven to destitution and constant bankruptcy. The other is the socialist road by which the peasants take their fate into their own hands under the leadership of the working class, and all become rich and prosperous together. The struggle over which of these two roads to take, is one that will exist for a long time during the period of transition in our country. However, as a result of the virtual completion of agricultural co-operation, the overwhelming majority of China's peasants have left the first road and are going along the second. The task ahead is to do our best to consolidate the system of co-operation while continuing to combat the spontaneous development of capitalism in the countryside.

Agricultural co-operation has opened up the broadest avenue for developing China's agricultural productive forces. Without agricultural co-operation, under the conditions of individual economy it would be

---

The English translation used in this study was published by the Foreign Language Press, Peking, in 1960. It should be noted that the program was presented as though wholly devised in 1956 and that the communes, which were established in 1958, were not envisioned in the program as presented.

mere idle fancy to strive for an average annual grain yield, with some exceptions, of 400, 500 and 800 catties per mou in the various major areas of the country within twelve years or for the aim of enabling, during the Second Five-Year Plan, the majority of the co-operatives to reach or exceed the output and income of the local well-to-do middle peasants working on their own. But with the coming of agricultural co-operation, together with the great achievements in socialist industrialisation under the First Five-Year Plan, and given the persistent efforts of all from now on, it is possible to attain the aims set out in this Programme.

The agricultural economy has great potentialities because of the generally good natural conditions in China, the abundance of rural manpower, fine traditions of industry and thrift among the peasants and their rich experience in intensive and meticulous farming. These potentialities of agriculture must be fully brought out on the basis of co-operation by every sort of vigorous and rational measure and by energetically extending agricultural mechanisation that is suited to actual conditions step by step and in a prepared manner. It also is necessary to combat conservatism in striving to attain the aims laid down in this Programme.

The raising of the level of agricultural production and the living standards of the peasants depends mainly on the hard work of the peasants themselves. However, the People's Government under the leadership of the working class and the Communist Party will of course do everything possible to help the peasants. The People's Government will from now on gradually give more and more of whatever assistance is necessary for carrying out the numerous measures to increase agricultural production as set out in this Programme. In fact, this means mutual support between the workers and peasants, the towns and the countryside.

The worker-peasant alliance led by the working class and the mutual support between the workers and peasants is the guarantee of the emancipation of the peasantry. The bourgeois rightist and feudal remnants have done their utmost to estrange relations between the workers and peasants and between the towns and countryside with the aim of restoring the landlord and capitalist systems. Such base attempts have failed and will continue to fail.

Education among the peasants should teach them to love their country, their co-operatives and their family as one. Without the People's Republic of China led by the Communist Party, the peasant masses would have continued to be ruled and exploited by the imperialists, landlords, rich peasants and speculating merchants, they could not possibly have had their own co-operatives and countless scenes of homelessness and death would have gone on being enacted. If one is to love one's own family, it is imperative to love the country and the co-operative. All forms of departmentalism and individualism which disregard the interests of the state and the collective interests of the co-operatives are wrong. In fact, they must end by jeopardising the interests of one's family.

Difficulties will still continue to emerge on the path of agricultural development. But man is the determining factor in doing things. To our liberated people, there are no difficulties that cannot be overcome. Fearlessness of difficulties is a great characteristic inherent in our working people.

This Programme is put forward for the whole nation. There are many different conditions in various localities and co-operatives. Therefore, on the basis of this Programme and in accordance with the specific conditions of each locality and co-operative, the leading organs of the Party and Government and the co-operatives in the various provinces (including municipalities and autonomous regions), special administrative regions (including autonomous chou), counties (including autonomous counties), districts, townships (including nationality townships) should realistically draw up their concrete plans for the development of the work in each locality, item by item and stage by stage, by means of the mass line. At the same time, all the state's economic departments, departments of science, culture, education and public health and the civic and judicial departments should also reexamine and renew their own working plans in accordance with the Programme.

Among the various tasks provided in this Programme, those such as afforestation, industry and thrift in running households, elimination of rats, flies and mosquitoes, the wiping out of the diseases from which the people suffer most seriously, and the advocation of birth control also apply to city residents. Such tasks can only be effectively carried out with the co-ordinated efforts of the towns and the countryside.

PROGRAM

1. Consolidate the agricultural co-operation system. Agricultural co-operation in our country was virtually completed in 1957. The future task is to strive to consolidate all the agricultural producers' co-operatives during the period of the Second Five-Year Plan, or a little longer.

The conditions for consolidating the co-operatives are: (1) To maintain the supremacy of the former poor peasants, hired farmhands and lower middle peasants (chiefly those who are now poor peasants, and the lower middle peasants among the new middle peasants) in the composition of the leadership of the co-operatives, and at the same time see to it that the upper middle peasants have appropriate representation. (2) To thoroughly carry out the policy of democratic management of the co-operatives. At fixed intervals the leading body of the co-operative should publicly declare the financial income and expenditure, and the functionaries should consult the masses on work and take part in production. (3) To thoroughly carry out the policy of running the co-operatives in the spirit of industry and thrift, and to fight against idleness and oppose extravagance and waste. (4) According to the economic conditions of the co-operatives and local natural conditions, to take various measures to increase production, gradually increase capital construction in agriculture, ensure adherence to and fulfillment of state plans, continuously expand production, and strive to enable the majority of the co-operatives to rise to or above the level of production and income of the local well-to-do middle peasants during the Second Five-Year Plan. (5) To handle the question of distribution rationally, give all-around consideration to the interests of the state, the co-operatives and the co-operative members; through the growth of production and in years of normal harvest, to increase the public accumulation funds of the co-operatives and the incomes of the co-operative members year by year so that the income of the collective economy of the co-operatives, plus the income from side-occupations undertaken by the co-operative members' families, will on an average per capita basis reach or surpass by around 1962 that of the local well-to-do middle peasants. (6) To strengthen political and ideological work, continuously raise the socialist consciousness of the co-operative members, overcome the capitalist ideology and departmental and individualist ways of thinking that run counter to the interests of the state and the collective interests of the co-operatives; in co-ordination with the rectification campaign among rural functionaries and the check-up in the co-operatives each year, systematically sum up the year's work and conduct intensive socialist education among the entire rural population. In multi-national areas, special attention should be paid to fostering solidarity and mutual help among the nationalities.

The few elementary co-operatives that still remain should be guided to switch over voluntarily to advanced co-operatives when the conditions are ripe. The education of and leadership over the small number of individual peasants who continue to work on their own should be strengthened, to enable them voluntarily and gradually to join the co-operatives. Those who will not join should be let alone.

2. Strive energetically for increased grain and other crops. In the twelve years starting from 1956, in areas north of the Yellow River, the Tsinling Mountains, the Pailung River and the Yellow River (in Chinghai), the average annual grain yield should be rised from the 1955 figure of over 150 catties per *mou* to 400 catties; in areas south of the Yellow River and north of the Huai River, the

yield should be raised from the 1955 figure of 208 catties to 500; in areas south of the Huai River, the Tsinling Mountains and the Pailung River, it should rise from the 1955 figure of 400 catties to 800 catties. In districts of sandy waste land or poor soil, in districts where there is waterlogging or drought all the year round, in mountainous districts at high altitude and with a cold climate, in districts with a very short period without frost, in sparsely populated districts and in districts where there is large-scale land reclamation, targets for increased production can be set with reference to specific conditions.

In the twelve years starting from 1956, the average annual yield of cotton (ginned) should be raised from the 1955 figure of 35 catties per mou (the average for the whole country) to 40, 60, 80 to 100 catties depending on local conditions.

While giving priority to increasing the output of grain, various places should develop a diversified agricultural economy, to ensure the attainment of the output figures set in the state plans for textile raw materials (cotton, bast fibre crops, silkworm cocoons), oil-bearing crops (soya beans, peanuts, rapeseed, sesame, tea-oil and tung-oil trees), sugar-bearing crops (sugar-cane and sugar-beet), tea, cured tobacco, fruits and medicinal herbs; and also take energetic measures to develop the cultivation of all other marketable industrial crops. In various provinces in South China where the conditions permit, attention should be paid to developing the cultivation of tropical and sub-tropical crops.

Agricultural co-operatives should encourage their members to grow vegetables and fodder crops on their own private plots. Co-operatives and state farms on the outskirts of cities and near industrial or mining districts should grow vegetables according to state plans so as to ensure a full supply of vegetables to the cities and industrial or mining districts.

3. Develop livestock breeding. In animal husbandry co-operation, different plans of development should be mapped out in accordance with local conditions. State livestock farms should be extended according to plan.

Great efforts should be made to protect and breed cattle, horses, donkeys, mules, camels, pigs, sheep, rabbits and other domestic animals and raise appropriate numbers of all types of poultry. Female, young and stud animals should be specially protected. Breeding stations should be set up, and the livestock strains improved.

In pastural areas, the pastures should be protected, and fodder grass grown and improved, and special attention should be paid to opening up sources of water. Livestock breeding co-operatives should gradually build up their own fields of fodder and grass, and silage should be popularised.

Within a period of from seven to twelve years, all possible areas should, in the main, have eliminated the most serious animal diseases such as rinderpest, hog cholera, Newcastle disease, contagious pleuropneumonia of cattle, foot and mouth disease, pork measles (cysticercosis), sheep pox, sheep mange, etc. By 1962, veterinary stations should have been set up in all townships in pastural areas. The co-operatives should have people who have had a basic training in the prevention and cure of animal diseases. The role of folk veterinarian should be developed fully, and they should be organised and guided to improve their technique and take part in the work of the prevention and cure of animal diseases.

4. Popularise measures to increase production and disseminate advanced experience—two fundamental conditions for increasing yields of farm crops. The chief measures to increase production are to: (1) Build water conservancy works; (2) Increase the supply of fertilisers; (3) Improve the old types of farm tools and extend the use of new types of farm tools; (4) Extend the use of good seed strains; (5) Expand the multiple-cropping areas; (6) Plant more high-yielding crops; (7) Carry out intensive and meticulous farming and improve farming methods; (8) Improve the soil; (9) Carry out water and soil conservation; (10) Protect and breed draught animals; (11) Wipe out insect pests and plant diseases; (12) Reclaim waste land and extend the cultivated area.

The chief steps in disseminating advanced experience are: (1) Provinces, municipalities and autonomous regions should collect local experiences in increasing yields and compile and publish books, so as to spread the knowledge; (2) Agricultural exhibitions should be held; (3) Conferences of model peasants should be called at regular intervals by the government at all levels, awarding and citing model peasants who have distinguished themselves in increasing production; (4) Mutual visits and comparisons should be arranged among the co-operatives to exchange experience in increasing production; (5) On the basis of summing up advanced experience, organisational work should be done to spread technical knowledge, and peasants and functionaries should be encouraged to study the advanced experience in management and technique of other co-operatives, other townships, counties and provinces (autonomous regions).

5. Build water conservancy works, expand irrigation, prevent and control flood and drought. In the twelve years starting from 1956, the development of water conservancy all over the country should be directed chiefly to the building of small and medium-sized water conservancy projects and, wherever necessary and possible, also to the building of large water conservancy projects.

In a planned way, as many small water conservancy projects as possible (such as digging wells and ponds, building dykes, sinking water storing wells, opening ditches, building embankments, reservoirs and a network of ditches, raising land and terracing land to store and drain water) should be built, and as many small rivers as possible should be harnessed, both types of projects being undertaken by the local governments and the agricultural co-operatives. Ordinary flood and drought should be eliminated, in the main, within twelve years by these measures that are to be co-ordinated with the building of medium-sized and large water conservancy projects and the harnessing of medium-sized and large rivers by the state.

In areas where waterlogging is serious, maximum efforts should be made to construct projects to wipe out waterlogging, drain water and transform low-lying land.

Paddy fields and irrigated land should be expanded in twelve years from more than 390 million mou in 1955 to approximately 900 million mou. While taking into account the varied conditions in the different localities, for the purpose of ensuring good harvests, the capacity of the irrigation facilities to deal with drought should be raised to between 30 and 50 days; and in places suitable for double-crop paddy fields, to between 50 and 70 days. To make full use of all available water resources, the departments concerned should work energetically to survey underground water and provide the water conservancy departments with the necessary data.

In building water conservancy projects where power-generation is possible, every effort should be made to undertake the simultaneous construction of small and medium-sized hydro-electric power stations; and by co-ordinating them with the medium-sized and large power stations built by the state, the supply of electricity to the countryside should be gradually increased.

6. Increase energetically the output of farm manure and chemical fertilizer. The agricultural co-operatives should undertake every measure to rely as far as possible on their own efforts for the supply of fertilizers. Particular attention should be paid to pig-breeding (in some places sheep-breeding). With the exception of certain national minority areas where pigs are not raised and of a small number of households which do not raise pigs because of religious beliefs, each household in the countryside should raise an average of one and a half to two pigs by 1962 and an average of two and a half to three pigs by 1967. Pigs should be kept in pigsties, sheep in sheepfolds, cattle and horses in barns. Energetic efforts should also be put into growing green manure crops according to local conditions and the fullest use should be made of night-soil from the cities and countryside and garbage and miscellaneous refuse which can be utilised as fertiliser.

Both the central and local authorities should work energetically to develop the chemical fertiliser industry and strive to produce approximately five to seven million tons by 1962 and approximately 15 million tons by 1967. Bacterial fertilizer should be increased to the fullest extent.

7. Improve old types of farm tools and extend the use of new types. Energetic efforts should be made to extend the use of improved and new types of farm tools suited to local conditions on the basis of the needs of production and by trial and remodelling to adapt them to the conditions of the particular region. Attention should always be paid to the maintenance and repair of farm tools. Technical guidance in the use of the new types of farm tools should be strengthened.

Keeping pace with the development of industrialisation of the country, mechanisation of agriculture should be carried out energetically step by step. Within the shortest possible time starting from 1956, the engineering and agricultural departments should draw up, on the basis of extensive experiments and research, a programme for agricultural mechanisation suited to the conditions in China, so as to make good preparations to extend the mechanisation of agriculture. Then machinery suited for use in various places should be manufactured for supply to the peasants; such machinery should be improved as circumstances require and its use extended energetically.

8. Put energetic efforts into propagating improved strains of crop seeds suited to local conditions and extending their use. Before 1962, extend the use of all sorts of existing improved strains of farm crops whose adaptability to local conditions has already been proved by selected experiments. In the case of improved strains (cotton for example) the use of which has in the main been popularised, steps should be taken to strengthen the work of invigorating strains and varying them. Major efforts should be made to cultivate new, improved strains and attention paid to trying out improved strains from other parts of the country and from abroad.

Agricultural co-operatives should set aside land specially for growing seed, strengthen the work of seed selecting among the masses and institute a system for propagating improved strains of seed and their alternate use. In areas where the harvest is unreliable, alternating between good and bad, attention should be paid to storing up good strains of seed. State farms under the central and local authorities should make themselves centres for propagating good strains of farm crops and take energetic steps to increase the available supplies and extend the use of improved strains of crops suited to local conditions. All provinces (including municipalities and autonomous regions), special administrative regions (including autonomous chou) and counties (including autonomous counties) should set up special seed agencies.

9. Extend the multiple-cropping area. In twelve years starting from 1956, the average multiple-crop index (including green manure crops), depending on the varying conditions in different areas, is to be raised to the following levels: (a) areas south of the Wuling Mountains, about 230 percent; (b) areas north of the Wuling Mountains and south of the Yangtse River, about 200 percent; (c) areas north of the Yangtse River and south of the Yellow River, the Tsinling Mountains and the Pailung River, about 160 percent; (d) areas north of the Yellow River, the Tsinling Mountains and the Pailung River and south of the Great Wall, about 120 percent; (e) in areas north of the Great Wall, in general, all cultivated land should be fully utilised and the area of unworked land reduced, and wherever possible energetic steps undertaken to expand the multiple-cropping areas.

10. Grow more high-yielding crops. All available water resources should be used to extend the area under rice. In the twelve years starting from 1956 the area under rice should be increased by 250 million mou. Such high-yielding crops as maize and tuber crops should be suitably developed according to the local needs and dietary habits of the people in the area.

11. Strive to improve methods of cultivation suited to local conditions. All agricultural co-operatives should undertake intensive and meticulous farming, improve their farming methods, and in a rational way apply fertiliser and irrigate their land, rotate crops, use inter-cropping, multiple cropping and close planting.

Field work should be done in good time. Sowing and weeding and thinning out in good time, improving field management, harvesting in good time and meticulously as well as careful threshing—all these things must be done to ensure good yields and good harvests.

12. Improve the soil. Agricultural co-operatives and state farms should actively improve and utilize saline and alkaline land, poor red soil and low-lying and sandy land and poor quality soil of other types. Attention should be paid to guarding against the land becoming saline and alkaline. Hilly land should be energetically terraced in a planned way. The utmost should be done to turn poor soil into rich, fertile land.

13. Extend water and soil conservation. In areas where water losses and soil erosion take place, the agricultural co-operatives should be relied upon and the masses mobilized on a wide scale, to undertake in a systematic and planned way water and soil conservation in co-ordination with their production work. In the twelve years starting from 1956, significant benefits from water and soil conservation must be reaped in all possible areas and water losses and soil erosion must be steadily reduced. In order to get quick results, the departments of agriculture, forestry, water conservancy, and animal husbandry, and scientific research institutions must co-operate

closely and give vigorous support to this kind of work under the unified leadership of the local Party and government organisations.

14. Protect and breed draught animals. In from seven to twelve years starting from 1956, every agricultural co-operative should raise an adequate number of strong draught animals. For this, sufficient grass and fodder crops must be produced and high-yielding fodder crops grown. The co-operatives must constantly improve the breeding management of draught animals, according to their conditions. Co-operatives and government authorities should take proper measures to encourage the breeding of draught animals.

15. Prevent, combat and eliminate plant diseases and insect pests. In from seven to twelve years starting from 1956, wherever possible, the insect pests and plant diseases that do the most harm to crops should be wiped out in the main, such as locusts, rice borers, army-worms, maize borers, aphids, red spiders, pink boll worms, Sitodiplosis mosellana Gehin, wheat smut, wheat nematode and black rot on sweet potato; at the same time the spread of other harmful insect pests, plant diseases and weeds should be prevented. All localities should include in their plans for destruction all other serious insect pests and plant diseases that can be wiped out. To this end, plant protection and quarantine should be strengthened.

The production of fungicides, insecticides and pest destroyers and their related apparatus should be developed in a planned way. The quality of such products and their supplies should be improved. Technical directions for their use should also be improved and their safety and effectiveness be ensured.

16. Reclaim waste land and extend the cultivated area. The state should reclaim waste land in a planned way. In the twelve years starting from 1956 the area cultivated by state farms should be increased from the 1955 figure of over 13 million mou to about 100 million mou. Wherever conditions permit, land reclamation should be carried out by organising new settlers and encouraging co-operatives to organise branch co-operatives or send out production brigades. The reclamation of waste land should be linked with the plans for water and soil conservation and livestock breeding, so as to prevent water losses, soil erosion, and damage to forests and indispensable grassland. Where forests have already been damaged, efforts should be made to restore them.

Agricultural co-operatives should make full use of scattered bits of idle land, such as ridges and corners of land plots, the edges of ponds, land along the ditches, and disused threshing grounds. According to conditions such land may be distributed to production brigades or co-operative members for private management.

In carrying out capital construction, the strictest economy in using land must be exercised by factories, mines, commercial and agricultural enterprises, departments of education, culture, public health, water conservancy, communications and military affairs, and mass organisations. They should refrain from using cultivated land, or otherwise occupy as little cultivated land as possible.

17. Develop the economy of mountainous areas. Agriculture, forestry, livestock breeding and production of various kinds of local and special products should be developed in the mountainous areas under the policy of promoting a diversified economy in a manner appropriate to local conditions.

The production of grain in mountainous areas should be developed, provided that this does not contradict the conservation of water and soil. In areas which are still short of grain, efforts should be made to achieve self-sufficiency or reduce the amount of grain that has to be imported from other areas.

In remote mountainous areas, emphasis should be laid on growing timber forests. In those that are less remote, the emphasis should be on growing forests of special economic value and for firewood and charcoal, on fruit growing and the development of local and special products. In mountainous areas in the south, attention should be paid to cultivating tea-oil and tung-oil trees, bamboo, mulberries and tea plants. In mountainous areas in the north, attention should be paid to growing walnuts, Mongolian oaks and wild pepper.

More medicinal herbs should be grown. Attention should be paid to protecting those which grow wild and, wherever possible, gradually bringing a certain amount under cultivation.

In the twelve years starting from 1956 all possible areas should by and large eliminate the animals which cause most damage to productive work in mountainous areas. Wild animals of economic value should be protected and their breeding encouraged.

Special administrative regions, counties, townships and agricultural co-operatives in mountainous and semi-mountainous areas should draw up production plans covering comparatively long periods. They should make rational use of the land and bring about an all-round development of production. All state departments concerned should properly co-ordinate and strongly support such work by organising communications and transport, supply and marketing, culture, education, public health work, and migration to these areas.

18. Develop forestry and clothe all possible bare land and mountains with green. In the twelve years starting from 1956, where natural conditions permit and where there is enough manpower to undertake the task, bare waste land and mountains should be clothed with green. Wherever possible trees should be planted in a planned way near houses, villages, and along roads and rivers. To achieve this, we must count on the agricultural co-operatives to plant trees and adhere to the policy that trees planted by co-operatives belong to

them. The co-operatives themselves should collect the seeds and grow saplings; there should be a division of labour with the due responsibility for work done, and those who cultivate the trees should ensure their growth. At the same time, co-operative members should be encouraged to grow trees near their own houses. The trees they cultivate should belong to them.

In addition to timber forests (including bamboo groves), the utmost use should be made of all manpower and uncultivated land in the cities and countryside to plant other trees of economic value such as fruit trees, mulberries, Mongolian oak, tea plants, varnish trees and oil-bearing trees.

Afforestation by the state should be vigorously promoted. State-owned forestry stations should mainly undertake large tracts of timber forests, while the creation of shelter belts for the conservation of water and soil, wind-breaks, sand-breaks and shelter belts along the sea coasts should also be undertaken in a planned way.

Departments of railways, communications and water conservancy, factories and mines, should plant and care for trees along railways, main roads and large rivers, around large reservoirs and mines, and enjoy the income derived from this source. Nearby agricultural co-operatives may also be entrusted with such afforestation work which should tally with the plans of government departments concerned; and the income derived from this source should accrue to the co-operatives.

Cultivated land should not be occupied for the purpose of afforestation.

Within twelve years state-owned forests should be put under proper management to the greatest extent possible. Small state-owned forests which the state finds inconvenient to manage itself should be entrusted to the care of the co-operatives. Forest resources should be protected and well cared for. Steps should be taken to prevent forest fires, prevent and deal with insect pests and plant diseases in forests; random felling and wasteful felling should be stopped. Timely steps should be taken to enable the re-growth of trees on the felling areas.

19. Raise the output of marine and fresh-water products and develop their breeding. In marine fishing, on the basis of co-operative organisation the latent potentialities of existing gear should be fully tapped and the fishing techniques gradually improved. Attention should be paid to increasing public accumulation funds, replenishing and improving fishing equipment, and gradually increasing the number of motor trawlers and motorised junks. Safety measures for production should be strengthened so as to extend fishing to deep seas. All possible water resources should be used to develop fresh-water fish farming. The work of breeding good stocks of fish and preventing and dealing with fish diseases should be strengthened. Energetic steps

should be taken to promote fish farming in shallow sea waters and the work to breed fish, shell-fish and aquatic grasses strengthened.

20. Run the state farms successfully. All state farms should unite with and help their neighbouring agricultural co-operatives and set good examples in farming technique as they are expected to do. They should develop a diversified economy, raise their efficiency in the use of manpower, continuously improve their production technique and labour management and raise their labour productivity. They should exercise the strictest economy, apply the principle of running the farms with industry and thrift, improve the wages system, and both raise their output and cut down costs of production.

21. Improve research in agricultural science and technical guidance. Wherever necessary and possible, organisations for research in agricultural science or to provide technical guidance should be set up or improved—such as academies of agricultural science, regional and specialised research institutes of agricultural science, provincial agricultural experimental stations, model county breeding farms and agricultural advisory centres—so that research in agricultural science may better serve the development of agriculture.

Scientific research in agriculture and technical guidance must be closely linked with the practical work of the peasant masses. Local peasants' experience in raising farm yields, particularly that of the old peasants, must be summed up and the useful farming experiences of other places must be studied. The staff of the county agricultural advisory centres should choose key farms and agricultural co-operatives as bases for their participation in farming.

To meet the needs of the growing co-operative economy, during the twelve years starting from 1956, technical personnel of primary and intermediate level should be trained for the agricultural co-operatives from among peasants who have certain farming experience and a certain level of general education.

22. Improve meteorological and hydrological work. In the main, the networks of meteorological observatories and stations and hydrological survey stations should be completed in the twelve years starting from 1956. Weather forecasting, in particular forecasting of hazardous weather, should be strengthened. Weather forecasting for agriculture should be established. All areas should give attention to meteorological broadcasts, so that precautions can be taken against natural calamities such as floods, droughts, gales and frost.

23. Run agricultural co-operatives and households in the spirit of industry and thrift. All agricultural co-operatives should apply the principle of industry and thrift in their work. Industry means giving full encouragement to members to work conscientiously and, wherever possible and necessary, to make vigorous efforts to branch out into new fields of production, develop a diversified economy and

exercise minute care in everything. Thrift means being strictly economical, lowering the costs of production and taking a firm stand against waste and extravagance. In all capital construction, co-operatives should make the fullest use of the manpower and material and financial resources at their disposal.

Education should be carried out among the peasants to encourage them to run their households in the spirit of industry and thrift, make long-term plans and arrange their household expenditure in a planned way. Women's organisations in the countryside should regard it as one of their important tasks to spread education among the women in running households in the spirit of industry and thrift and doing a good job of housework. Attention should be given to popularising model examples of peasant household management in which every item is carefully counted, and thrift is exercised in every outlay, and saving and accumulation are increased. It is a wrong idea to make no long-term plans, to disregard household accumulation and to consume everything available at any time.

Expenditure on weddings, funerals and social occasions should be cut down in the countryside, and irrational past customs and habits changed.

24. Make fuller use of manpower, raise labour productivity and develop diversified economy in the agricultural co-operatives. Agricultural co-operatives must make fuller use of manpower in order to fully develop arable farming and side-occupations like forestry, livestock breeding and fishery, together with the rural handicrafts, and seek out fresh sources of production so as to increase the wealth of society and the income of the rural population. In the north, means should be found to undertake more forms of work during winter. Energetic efforts should be made at the same time to improve technical skills in production, improve labour organisation and management and steadily raise the labour productivity of the members of the co-operatives.

In the seven years starting from 1956, every able-bodied man in the countryside should be able to put in at least about 250 working days a year. The labour power of woman should be reasonably arranged and organised with consideration given to their particular characteristics. Within the seven years and depending on the different conditions in the various areas, every able-bodied woman in the countryside should, apart from time spent on household work, be able to give no less than 80 to 180 days a year, to agriculture or side-occupations (including household side-occupations). In addition, the co-operatives should make appropriate arrangements for all those in the countryside who can undertake only part-time work or who are fitted for light work, and encourage them to take an active part in whatever work they are fit for.

25. Build up grain reserves. In the twelve years from 1956, all agricultural co-operatives except those which engage mainly in mountain forestry or growing industrial crops and are short of grain, should make concrete plans in accordance with their own conditions to build up a surplus of grain against times of urgent need that, added to that of the members' families, is sufficient for three, six, twelve or eighteen months' consumption. Where harvests are habitually unreliable and communications inconvenient, special attention should be paid to making up for poor harvests in the years of rich harvest and to accumulating grain against famine.

In the twelve years from 1956, the state should build up enough grain stocks for one to two years' consumption, so as to meet urgent needs.

26. Improve housing conditions. As their production grows and their members' incomes increase, the agricultural co-operatives should encourage and assist the members to repair and build dwelling houses and thus improve their housing conditions, according to the needs and possibilities. This should be done in a prepared, planned way, at different times and group by group, under the principle of voluntariness, mutual help, and economical use of money and land.

27. Wipe out the four pests. In the twelve years beginning from 1956, rats, bedbugs, flies and mosquitoes should be by and large wiped out wherever possible.

28. Strive to eliminate the most serious diseases among the people. In the twelve years beginning from 1956 the diseases from which the people suffer most seriously should be virtually wiped out wherever possible. These include schistosomiasis, smallpox, bubonic plague, malaria, kala-azar, hookworm, filariasis, tetanus neonatorum, and venereal diseases. Energetic measures should also be taken to prevent and cure other diseases such as measles, dysentery, typhoid fever, Japanese Type-B encephalitis, poliomyelitis, diphtheria, pulmonary tuberculosis, leprosy, trachoma, goitre, Kaschin-Beck's disease, and Keshan disease. There should be vigorous efforts to train medical personnel, including traditional Chinese doctors.

Energetic efforts should be made to unfold a regular, patriotic, mass movement for public health, so that every person and family acquire good habits of cleanliness and hygiene. The underlying spirit of paying attention to cleanliness and hygiene is to eliminate diseases, bring about a general invigoration, reform the outmoded habits and customs and transform the country.

29. Protect women and children. The principle of equal pay for equal work must be resolutely put into practice in relation to women's productive work. Whenever necessary and possible, agricultural co-operatives may set up temporary, simple creches suitable for the

busy farming periods. In allocating work the physiological character of women members must be taken into consideration. Suitable regulations should be made concerning the participation of children in the countryside in auxiliary work during busy seasons, taking their age and physical strength into consideration.

Public health departments should train midwives for the rural areas, exert their utmost efforts to popularise modern methods of delivery, provide post-natal and infant care and cut down the incidence of maternal diseases and the infant mortality rate.

Other than in areas inhabited by national minority peoples, birth control should be publicised and popularised in all densely populated areas and family planning advocated, so that the family can avoid being overburdened and a better education can be given to the children and full chances of employment provided.

30. Carry out the "five guarantees"; care for the dependants of revolutionary martyrs and the disabled revolutionary ex-service men; support and respect parents. As regards those members—widows, widowers, orphans and the childless—who lack labour power and are without means of support, the co-operative should make overall plans and assign production brigades or teams to make proper arrangements in production for them so that they can take part in work they are physically able to do; and to give due consideration to their livelihood so that all will be guaranteed a regular supply of food, clothing and fuel, the young a chance of education and the aged a proper burial after death. In a word, they will be guaranteed a means of livelihood in their lifetime and a proper burial after death.

In accordance with the state regulations, the co-operatives should give favourable treatment to the dependants of revolutionary martyrs who lack labour power and to disabled revolutionary ex-service men who cannot make a living even after receiving the ex-service men's pensions. Their living standards should be no lower than that of the average for the co-operative.

Young and middle-aged men and women should be taught that they should support and respect their parents so that the aged who are incapable of working will receive reasonable care during their lives and enjoy mental ease.

31. Wipe out illiteracy, develop cultural and educational work in the rural areas. In the twelve years beginning from 1956, depending on the local conditions, illiteracy among young and middle-aged people must be virtually wiped out. Efforts should be made to set up spare-time schools step by step in the townships or co-operatives to further enhance the educational level of the functionaries in the primary rural units and the peasants. Diversified methods should be adopted in setting up rural schools. In order that primary education can gradually become universal, apart from schools established by

the state, energetic steps must be taken to encourage people to set up schools collectively and the setting up of private schools should be allowed.

Following the development of agricultural production, the co-operatives should gradually improve and develop cultural and recreational work as conditions permit and in accordance with the principle of industry and thrift in building the country, in running the co-operatives and in managing households.

32. Develop the radio diffusion network in the rural areas. In the seven or twelve years starting from 1956, depending on local circumstances, the radio diffusion network is to be extended to practically all rural areas. The majority of the producers' co-operatives in agriculture, forestry, fishery, livestock breeding, salt production and handicrafts should be enabled to receive the radio programmes.

33. Develop the rural telephone and postal network. In the seven or twelve years starting from 1956, depending on local conditions, a telephone network should be set up in the townships and in a certain number of co-operatives. Radio telephone equipment is to be installed in some areas. Within the twelve years, a postal network should cover virtually all rural areas so that there are good postal and telegraphic services.

34. Develop rural communications and transport. Within the seven or twelve years starting from 1956, depending on local conditions, a network of local roads should be built throughout practically the entire country. Roads of various kinds suited to local means of transport should gradually be built by relying on the efforts of the masses, in accordance with growing local requirements for transport services between provinces (including municipalities or autonomous regions), between special administrative regions (including autonomous chou), between counties (including autonomous counties), between districts and between townships (including nationality townships). The following principles should govern road construction: realistic consideration of the actual conditions, suitability to local circumstances, use of local materials, economy and practicality in use, and no waste of land. At the same time, all roads must be well maintained. Special attention must be paid to road construction in the mountainous areas.

In places served by rivers, navigable channels, if possible, should be dredged and kept in good order to improve communications.

35. Adjust the rural network of commerce. In order to meet the new conditions arising after agricultural co-operation, the commercial departments and the supply and marketing co-operatives in the rural areas should complete the adjustment of the rural commercial network in the next few years. They should speed up the

circulation of goods, strengthen the administration of primary markets, ensure that all rural areas get good service in the supply of goods and purchase of agricultural produce, combat the spontaneous development of capitalism and facilitate the development of rural production.

36. Promote rural credit co-operation. The rural credit co-operatives must be consolidated and energetic efforts made to extend rural credit and savings services, so as to help the agricultural co-operatives and peasants meet their needs for short-term capital turn-over.

37. Make full use of the initiative of ex-service men in building socialism in the countryside. Ex-service men in the rural areas should maintain the fine traditions of the People's Liberation Army on the front of socialist production. They should be hard-working and plain-living, maintain close contacts with the masses, abide by the state's policies, laws and regulations, and labour discipline and play an active role in agricultural production. The township people's councils and the agricultural co-operatives must make suitable arrangements for ex-service men to take part in production, and help them learn and improve their skill in production. The people's governments at various levels must pay constant attention to the progress of the ex-service men and give them political and ideological education.

38. Raise the socialist initiative of the youth in the countryside. The young people in the rural areas should be educated to ardently love their motherland, their villages and labour, and love their co-operatives as dearly as their homes. They must be encouraged to exert themselves in learning to read and write, studying agricultural science and farming techniques and learning from the production experience of the middle-aged and old peasants. The rural youth should become a shock force in production, construction and science and culture in the countryside.

Those with a primary or secondary school education in the cities, other than those who can enter higher schools or find local jobs, should enthusiastically respond to the state's call, go to the villages and mountainous areas and take part in agricultural production and in the great cause of socialist agricultural construction. Eighty-five percent of the population of our country are in the countryside and industry cannot develop by itself without the development of agriculture. To work in the rural areas is both an urgent necessity and a great honour.

39. Remould the former landlords, rich peasants, counter-revolutionaries and other bad elements and safeguard socialist order in the countryside. Former landlords and the rich peasants who have given up exploitation and former counter-revolutionaries in the rural areas may be admitted individually into the agricultural co-operatives as members or candidate members in accordance with the "Model

Regulations for Advanced Agricultural Producers' Co-operatives" and on consideration of their actual behaviour. Those who are not qualified for admission may be referred to the co-operatives by the township people's council and work under the co-operative's supervision. The co-operatives must make greater efforts to educate and control them according to individual cases; and must constantly enhance the vigilance of the co-operative members and the peasants outside the co-operatives against any subversive activities they may attempt. Former landlords, rich peasants and counter-revolutionaries who have already become members or candidate members of the co-operatives, if they persistently misbehave despite repeated instructions, may be demoted to candidate members or to the status of working under supervision in the case of members, and to the status of working under supervision in the case of candidate members. Those who commit sabotage should be punished according to law.

Gambling is strictly prohibited and the activities of secret and superstitious societies are banned. Thieves, swindlers, hooligans, special agents and various bad elements seriously disrupting social order should be punished according to law.

40. Workers in the cities and peasants in the co-operatives must support each other. The workers should turn out more and better industrial goods to meet the peasants' needs; the peasants should grow more and better grain and industrial raw materials to meet the needs of industry and the town dwellers. Workers in the cities and peasants in the co-operatives should also arrange get-togethers, visit one another, and exchange correspondence. Through ways like these, they should strengthen their contacts, give each other encouragement and swap experience so as to further the development of industry and agriculture and the consolidation of the worker-peasant alliance led by the working class.

INDEX

UNIVERSITY
OF BRISTOL
LIBRARY

ABOUT THE AUTHOR

LESLIE T. C. KUO served for 11 years as a senior officer of the Food and Agriculture Organization (FAO) of the United Nations. Dr. Kuo was appointed chief of the Oriental Project, National Agricultural Library, U. S. Department of Agriculture, in 1960. He did his undergraduate work in economics at Nankai University, Tientsin, China, and obtained his master's degrees (agricultural economics and regional and city planning) and Ph. D. (agricultural economics) at Cornell University.

Dr. Kuo has written numerous articles for professional publications, including the Library Journal, the China Quarterly (London), the International Development Review (Washington, D. C.), Focus (American Geographical Society, New York), and publications by the American Association for the Advancement of Science, as well as reports and documents for FAO. He also has supervised the preparation of a series of bibliographies on Chinese, Japanese, and Korean agriculture, published by the National Agricultural Library.

Dr. Kuo visited the PRC in the summer of 1975, during which trip he visited several communes and handicraft factories.

RELATED TITLES
Published by
Praeger Special Studies

THE AGRICULTURAL DEVELOPMENT OF IRAN
Oddvar Aresvik

THE AGRICULTURAL DEVELOPMENT OF JORDAN
Oddvar Aresvik

THE AGRICULTURAL DEVELOPMENT OF TURKEY
Oddvar Aresvik

CHINA'S ENERGY: Achievements, Problems, Prospects
Vaclav Smil

CHINA'S PETROLEUM INDUSTRY: Output Growth and Export Potential
Chu-yuan Cheng

AGRICULTURAL SUPPLY RESPONSE: A Survey of the Econometric Evidence
Hossein Askari
John Thomas Cummings

THE ECONOMICS OF THE GREEN REVOLUTION IN PAKISTAN
Mahmood Hasan Khan

FACTIONAL AND COALITION POLITICS IN CHINA: The Rebuilding of the Communist Party
Y. C. Chang

77 03233

UNIVERSITY OF BRADFORD
Library

Accession No. 0340735070

This book should be returned not later than the last date stamped below. The loan period may be extended on request provided there is no waiting list.
FINES ARE CHARGED ON OVERDUE BOOKS

ONE WEEK ONLY

NOT FOR LOAN UNTIL 16 FEB 1995 UNIVERSITY OF BRADFORD LIBRARY

22 APR 1997

23 APR 1996

29 APR 1996

L45 14 JAN 2003 350

BD 034073507 0